SpringerBriefs in Water Science and Technology

W0016676

More information about this series at http://www.springer.com/series/11214

Samiha Ouda

Major Crops and Water Scarcity in Egypt

Irrigation Water Management under Changing Climate

 Springer

Tahany Noreldin
Mohamed Hosney
Khaled Abd El-Latif
Fouad Khalil
Abd El-Hafeez Zohry
Ahmed M. Taha
Mahmoud A. Mahmoud
Sayed abd El-Hafez
Ahmed Said
A.Z. El-Bably

Agricultural Research Center
Giza
Egypt

ISSN 2194-7244   ISSN 2194-7252   (electronic)
SpringerBriefs in Water Science and Technology
ISBN 978-3-319-21770-3   ISBN 978-3-319-21771-0   (eBook)
DOI 10.1007/978-3-319-21771-0

Library of Congress Control Number: 2015945154

Springer Cham Heidelberg New York Dordrecht London

Printed on acid-free paper

Springer International Publishing AG Switzerland is part of Springer Science+Business Media
(www.springer.com)

Contents

Introduction

Egypt is located in the eastern corner of Africa on the Mediterranean Sea. The agricultural area in Egypt is composed of two parts: Nile Delta and Valley, which is the main contributor to food production, trading activities and national economy. It is also the most densely populated area in Egypt. Climate variability and change have forced us to think about the future of water resources and its sustainability under the current situation of water scarcity. The limited share of the Nile water that Egypt receives is not expected to increase in the future. Taking into account the population growth and the expected negative effect of climate change on rain in Ethiopia, Egypt will face a problem to allocate water to agriculture to maintain food security. The major crops involve in food security in Egypt are: wheat, maize, rice and sugarcane. There is a gap between production and consumption of wheat, maize and sugarcane. Whereas, for rice, there is a need to reduce its cultivated area, as its water consumption is high and its production is enough for local consumption at the time being.

The actual water resources currently available for use in Egypt are 55.5 BCM/year, and 1.3 BCM/year effective rainfalls on the northern strip of the Nile Delta, non-renewable groundwater for western desert and Sinai, while water requirements for different sectors are of the order of 79.5 BCM/year. The gap between the needs and availability of water is about 20 BCM/year. This gap is overcome by recycling agricultural drainage water. Egypt has reached a state where the quantity of water available is imposing limits on its national economic development. As indication of scarcity in absolute terms, often the threshold value of 1000 m^3/capita/year is used. Egypt has passed that threshold already. As a threshold of absolute scarcity 500 m^3/capita/year is used, this will be evident with population predictions for 2025, which will bring Egypt down to 500 m^3/capita/year (Ministry of Irrigation and Water Resources, 2014). Since 85 % of the total available water is consumed in agriculture and most of the on-farm irrigation systems are low efficient coupled with poor irrigation management, water scarcity will negatively affect food security. Furthermore, surface irrigation is the major system in Egypt applied to 83 % of the old cultivated land (Nile Delta and Valley). Application efficiency of surface irrigation in Egypt is 60 %, which endures large losses in the applied irrigation water to drainage canals.

Previous research on the effect of using improved agricultural management practices on cultivated crops revealed that cultivation on raised beds could reduce the applied irrigation water by 20 %, which result in yield increase by 15 % (Abouelenein et al. 2009). Furthermore, changing application efficiency from 60 % under surface irrigation to 80 % under sprinkler system or 95 % under drip irrigation could save large amounts of irrigation water. Under this practice, yield is expected to increase by 15 and 18 % under sprinkler and drip system, respectively (Taha 2012). Thus, existing agricultural water management technologies are available to help meet the challenge of water scarcity. Furthermore, the saved amounts of irrigation water can be used to cultivate new land and increase national production of these crops.

The uncertain climate change impacts on the Nile flow could add another challenge for water management in Egypt. Studies on the effect of climate change on the Nile flow clearly show that the assessment is strongly dependent on the choice of the climate scenario and the underlying GCM model. For temperature, although the magnitude of the change varies, the direction of change is clear; all models expect temperatures to rise. For rainfall, however, not only the magnitude varies substantially across the models, but even the signal of the change varies. The choice of the emission scenario also leads to different estimates (Sayed 2004). There are large uncertainties attached at all the steps of scenario construction that need to be quantified in the analysis of future impacts. In addition, the reviewed studies show that the Nile flow is extremely sensitive to climate, and especially rainfall changes due to the highly nonlinear relationship between precipitation and runoff (Sayed 2004). Thus, water gap in Egypt could increase in the future and under expected climate change (Sanchez et al. 2005).

As reported by Eid (2001), a temperature rise by 1 °C may increase evapotranspiration (ET) rate by about 4–5 %, while a rise of 3 °C may increase ET rate by about 15 %. Furthermore, Attaher et al. (2006) and Khalil (2013) concluded that the future climate change will increase potential irrigation demands, due to the increase in evapotranspiration (ET) in 2100. While Ouda et al. (2011) developed prediction equations to calculate total water requirements needed to support irrigation in Egypt in 2025 and they found that an increase by 33 % in water required for irrigation is expected to occur as a result of temperature increase by 2 °C and population growth. Thus, crop production in Egypt will be highly vulnerable to climate change due to increase in its water requirements that will reduce cultivated area and consequently reduce total production.

Our objectives were to quantify the effect of climate change on production of wheat, maize, rice and sugarcane cultivated in the governorates of Nile Delta and Valley. Furthermore, quantification of the effect of strategies could be used as adaptation to reduce the risk of climate change on these crops. These quantifications are very important for policy makers to be included in their future plans.

The studied area is composed of 17 governorates in the Nile Delta and Valley. These governorates are: Alexandria (latitude 31.70°, longitude 29.00° and elevation

7.00 m), Demiatte (latitude 31.25°, longitude 31.49° and elevation 5.00 m), Kafr El-Sheik (latitude 31.07°, longitude 30.57° and elevation 20.00 m), El-Dakahlia (Latitude 31.03°, longitude 31.23° and elevation 7.00 m), El-Behira (latitude 31.02°, longitude 30.28° and elevation 6.70 m), El-Gharbia (latitude 30.47°, longitude 32.14° and elevation 14.80 m), El-Monofia (latitude 30.36°, longitude 31.01° and elevation 17.90 m), El-Sharkia (latitude 30.35°, longitude 31.30° and elevation 13.00 m), El-Kalubia (latitude 30.28°, longitude 31.11° and elevation 14.00 m), El-Giza (latitude 30.02°, longitude 31.13° and elevation 22.50 m), El-Fayoum (latitude 29.18°, longitude 30.51° and elevation 30.00 m), Beni Swief (latitude 29.04°, longitude 31.06° and elevation 30.40 m), El-Minia (latitude 28.05°, longitude 30.44° and elevation 40.00 m), Assuit (latitude 27.11°, longitude 31.06° and elevation 71.00 m) Suhag (latitude 26.36°, longitude 31.38° and elevation 68.70 m), Qena (latitude 26.10°, longitude 32.43° and elevation 72.60 m) and Aswan (latitude 24.02°, longitude 32.53° and elevation 108.30 m).

Alexandria is the wettest city in Egypt. Demiatte and Kafr El-Sheik are governorates with the least temperature fluctuation between day and night. Minia, Assuit, Qena and Suhag are governorates with the most temperature fluctuation between day and night. Aswan is the hottest in summer days. Figure 1 shows the map of Nile Delta and Valley governorates.

To accomplish the quantification of the effect of climate change on production of the selected crops, ET, crop factor and water requirements for the selected crops should be calculated under present time. Furthermore, a similar procedure should be

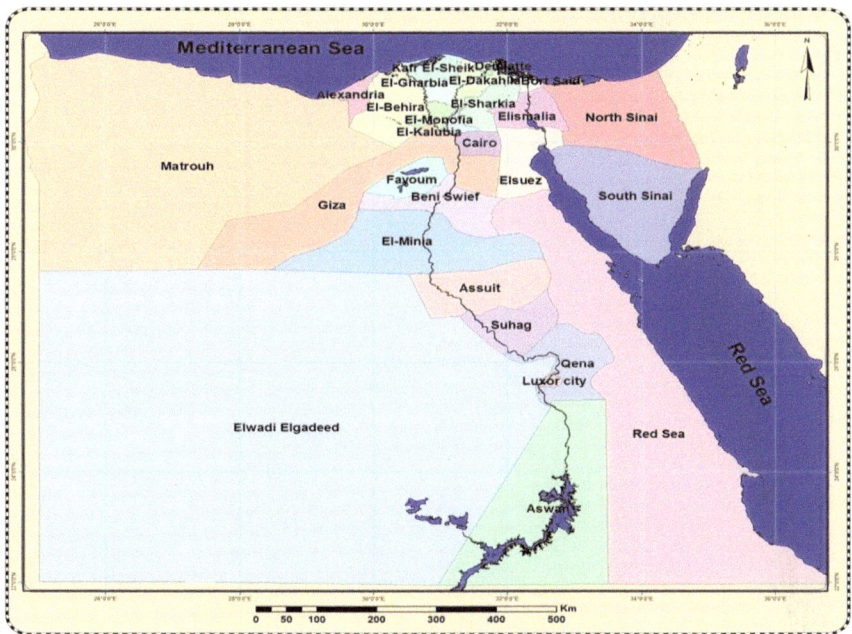

Fig. 1 Map of Nile Delta and Valley of Egypt

carried out under the future climate change scenario developed by an Atmospheric Oceanic General Circulation model to determine the percentage of increase in water requirements for each of the selected crops. The effect of improved water management practices, such as cultivation on raised beds or increasing water application efficiency on the selected crops will be investigated under both present time and climate change, where it can be used as adaptations to climate change.

References

Abouelenein R, Oweis T, El Sherif M, Awad H, Foaad F, Abd El Hafez S, Hammam A, Karajeh F, Karo M, Linda A (2009) Improving wheat water productivity under different methods of irrigation management and nitrogen fertilizer rates. Egypt J Appl Sci 24(12A):417–431

Attaher S, Medany M, AbdelAziz AA, El-Gendi A (2006) Irrigation-water demands under current and future climate conditions in Egypt. The 14th annual conference of the misr society of agricultural. engineering 1051–1063

Eid H (2001) Climate change studies on Egyptian Agriculture, Soils, Water and Environment research institute SWERI ARC, Ministry of Agriculture, Giza, Egypt.

Khalil AA (2013) Effect of climate change on evapotranspiration in Egypt. Researcher 51:7–12

Ministry of Irrigation and Water Resources (2014) Water scarcity in Egypt: The urgent need for regional cooperation among the Nile Basin countries. Technical report

Ouda S, Khalil F, El Afendi G, Abd El-Hafez S (2011) Prediction of total water requirements for agriculture in the Arab World under climate change. In: 15th international water technology conference 1150–1163

Sanchez P, Swaminathan MS, Dobie P, Yuksel N (2005) Halving hunger: it can be done, UN millennium project, UNDP

Sayed MAA (2004) Impacts of climate change on the Nile Flows, Ain Shams University, Cairo, Egypt

Taha A (2012) Effect of climate change on maize and wheat grown under fertigation treatments in newly reclaimed soil. Ph.D. Thesis, Tanta University, Egypt

Chapter 1
Evapotranspiration Under Changing Climate

Samiha Ouda, Tahany Noreldin and Mohamed Hosney

Abstract This chapter described methodology to calculate evapotranspiration (ET) values similar to the values calculated with Penman–Monteith equation (P–M), using ET values calculated by Hargreaves–Samani equation (H–S) under current and climate change. The BISm model was used to calculate monthly values of ET using P–M and H–S equations using weather data averaged over 10 years, from 2004 to 2013 for each of the 17 studied governorates and the values were compared. The comparison showed that there were deviations between monthly ET values calculated for each equation in each governorate. Thus, a linear regression equation was established with ET values resulted from P–M plotted as the dependent variable and ET values from H–S equation plotted as the independent variable. The quality of the fit between the two methodologies was presented in terms of the coefficient of determination (R^2) and root mean square error per observation (RMSE/obs). ECHAM5 climate change model was used to develop A1B climate change scenario for each governorate for the years 2020, 2030 and 2040, where ET values were calculated. The results indicated that R^2 was between close to one and RMSE/obs values were close to zero. The results also indicated that the calibration coefficients were capable to account for the effect of relative humidity, wind speed and potential sunshine hours, which were not included in the H–S equation. Furthermore, under A1B climate change scenario, the values of ET were increased. The above methodology could solve a large problem that faces researchers and extension workers in irrigation scheduling in Egypt and in other developing countries under current climate and in calculation of water requirements under climate change.

Keywords Penman–Monteith and Hargreaves–Samani equations · BISm model · ECHAM5 model · A1B climate change scenario

Climate plays an important role in crop production. Crops growth periods, crops water requirements, and scheduling irrigation for crops are dependent on weather conditions. The calculation of the evapotranspiration (ET) includes all the weather

S. Ouda (✉) · T. Noreldin · M. Hosney
Water Requirements and Field Irrigation Research Department, Agricultural Research Center,
Soils, Water and Environment Research Institute, El Giza, Egypt

© The Author(s) 2016
S. Ouda, *Major Crops and Water Scarcity in Egypt*,
SpringerBriefs in Water Science and Technology,
DOI 10.1007/978-3-319-21771-0_1

1

parameters prevailed in a specific area. ET is a combination of two processes: water evaporation from soil surface and transpiration from the growing plants (Gardner et al. 1985). Direct solar radiation and, to a lesser extent, the ambient temperature of the air provide energy for evaporation, whereas solar radiation, air temperature, air humidity, and wind speed should be considered when assessing transpiration (Allen et al. 1998). ET is a key component in hydrological studies. It is used for agricultural and urban planning, irrigation scheduling, regional water balance studies, and agro-climatic zoning (Khalil et al. 2011).

Various equations are available for estimating ET. These equations range from the most complex energy balance equations requiring detailed climatological data (Penman–Monteith; Allen et al. 1989) to simpler equations requiring limited data (Blaney–Criddle 1950; Hargreaves–Samani 1982, 1985). The Penman–Monteith equation (P–M) is widely recommended because of its detailed theoretical base and its accommodation of small time periods. The method requires maximum and minimum temperature, relative humidity, wind speed, and potential sunshine hours, whereas Hargreaves–Samani equation (H–S) requires three weather parameters only: maximum and minimum temperatures and solar radiation.

This chapter described methodology to calculate ET values similar to the values calculated with P–M equation, using ET values calculated by H–S equation.

BISm Model Description

The Basic Irrigation Scheduling model (BISm) was written using MS Excel to help people plan irrigation management of crops. The BISm model calculates ET using the Penman–Monteith (P–M) equation (Monteith 1965) as presented in the United Nations FAO Irrigation and Drainage Paper (FAO, 56) by Allen et al. (1998). If only temperature and solar radiation data are input, Hargreaves–Samani equation is used (Snyder et al. 2004). The weather station latitude and elevation must also be input. After calculating daily means by month, a cubic spline curve fitting subroutine is used to estimate daily ET rates for the entire year.

The BISm model was used to calculate monthly ET values as an average over 10 years, from 2004 to 2013 for each of the 17 studied governorates using P–M equation. Furthermore, ET values using H–S equation for these governorates were calculated and then compared to ET values of P–M equation.

Comparison Between ET(P–M) and ET(H–S) Values

The calculated values of ET(P–M) and ET(H–S) in each governorate were graphed to ease comparison. The results for Alexandria (Fig. 1.1), Demiatte (Fig. 1.2), Kafr El-Sheik (Fig. 1.3), and El-Dakahlia (Fig. 1.4) showed that during summer months, the H–S equation underestimated ET values.

Fig. 1.1 Comparison between ET(P–M) and ET(H–S) in Alexandria governorate

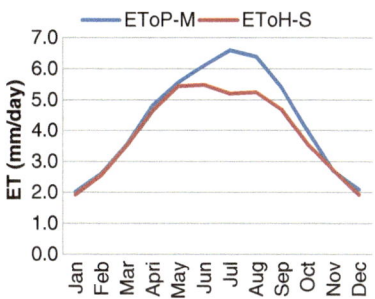

Fig. 1.2 Comparison between ET(P–M) and ET(H–S) in Demiatte governorate

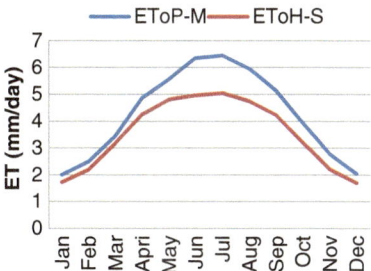

Fig. 1.3 Comparison between ET(P–M) and ET(H–S) in Kafr El-Sheik governorate

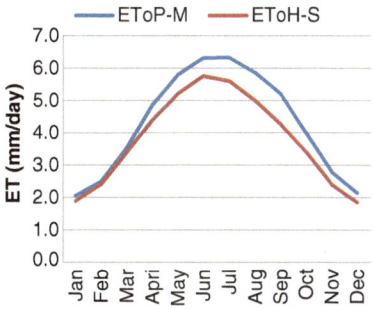

Fig. 1.4 Comparison between ET(P–M) and ET(H–S) in El-Dakahlia governorate

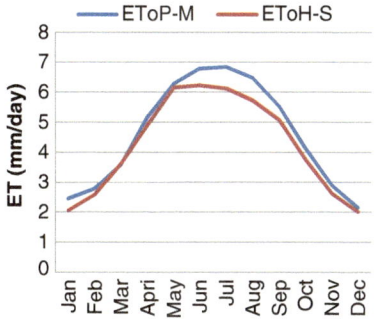

Regarding to El-Behira and El-Gharbia, the deviation of ET(H–S) from ET(P–M) was high during all the year, especially in the summer months (Figs. 1.5 and 1.6). Similarly, Figs. (1.7 and 1.8) show the same trend in Assuit and Aswan governorates. However, the deviation in the winter months became higher.

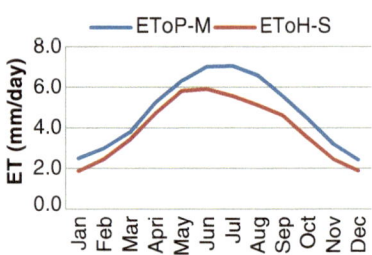

Fig. 1.5 Comparison between ET(P–M) and ET(H–S) in El-Behira governorate

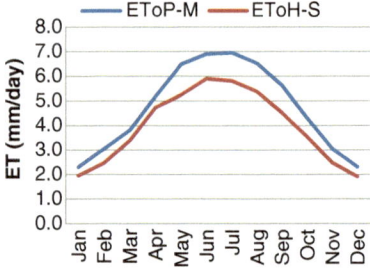

Fig. 1.6 Comparison between ET(P–M) and ET(H–S) in El-Gharbia governorate

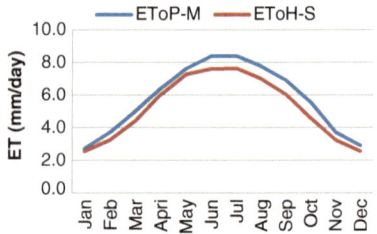

Fig. 1.7 Comparison between ET(P–M) and ET(H–S) in Assuit governorate

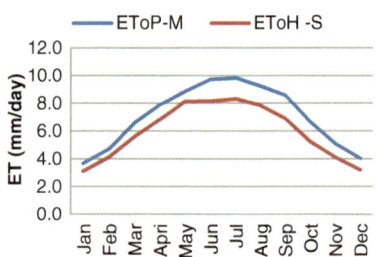

Fig. 1.8 Comparison between ET(P–M) and ET(H–S) in Aswan governorate

With respect to El-Monofia, Fig. (1.9), and El-Sharkia (Fig. 1.10), the values ET (P–M) and ET(H–S) were close to each other from January to May, and then the deviations were higher in the months after.

In El-Kalubia governorate (Fig. 1.11), the situation was different, where low deviation was observed only in the summer months, whereas there was no deviation between the values of ET(P–M) and ET(H–S) for the rest of the months. Similar trend was observed in Beni Swief governorate with higher deviation from May to August (Fig. 1.12). In El-Minia governorate, the deviation was very low (Fig. 1.13), and in Suhag governorate (Fig. 1.14), the deviation was higher from May to July.

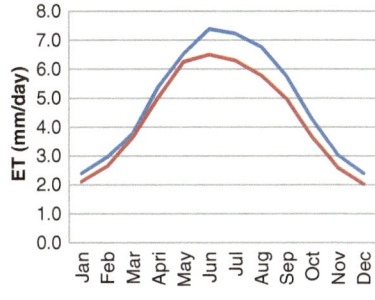

Fig. 1.9 Comparison between ET(P–M) and ET(H–S) in El-Monofia governorate

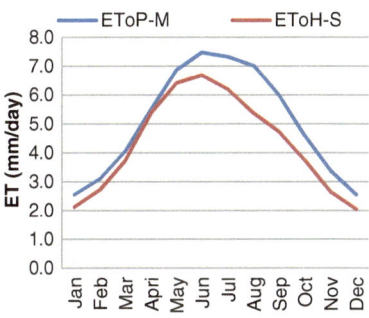

Fig. 1.10 Comparison between ET(P–M) and ET(H–S) in El-Sharkia governorate

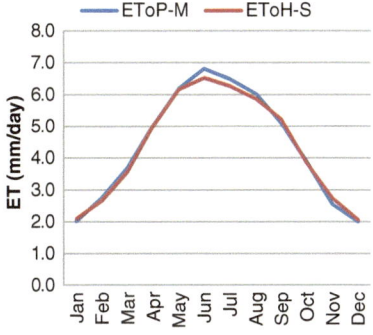

Fig. 1.11 Comparison between ET(P–M) and ET(H–S) in El-Kalubia governorate

Fig. 1.12 Comparison between ET(P–M) and ET(H–S) in Beni Swief governorate

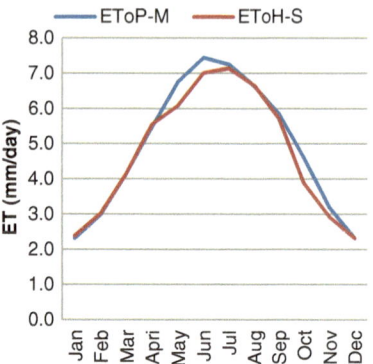

Fig. 1.13 Comparison between ET(P–M) and ET(H–S) in El-Minia governorate

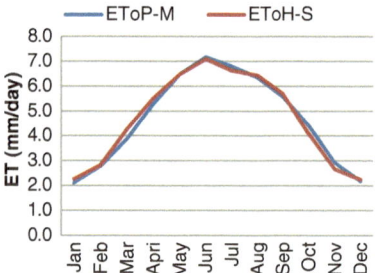

Fig. 1.14 Comparison between ET(P–M) and ET(H–S) in Suhag governorate

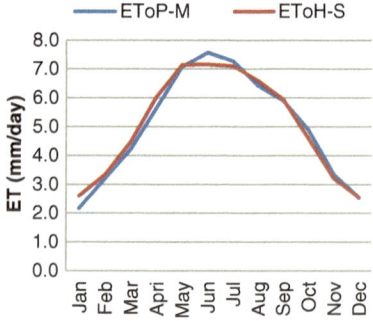

The difference between ET(P–M) and ET(H–S) for El-Giza (Fig. 1.15) and El-Fayoum (Fig. 1.16) was low for most of the months.

Regarding to Qena governorate, there was no difference between values of ET (P–M) and ET(H–S) in June and July. However, for the rest of the months, there were no differences (Fig. 1.17).

The above comparison showed that there were deviations between monthly ET values calculated for each equation in each governorate. Therefore, to increase the accuracy of the estimation, a linear regression equation was established with ET values resulted from P–M plotted as the dependent variable and ET values from H–S

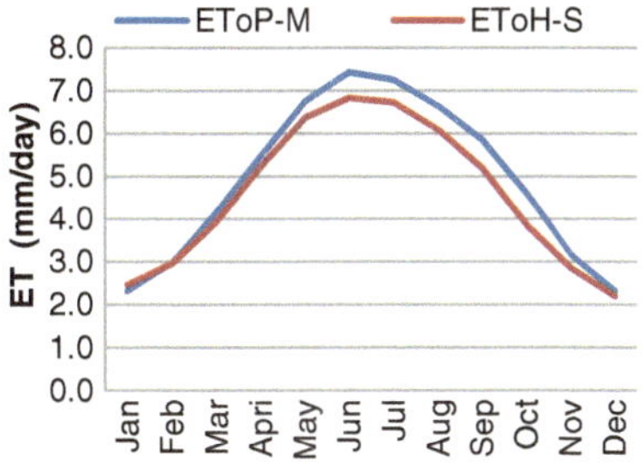

Fig. 1.15 Comparison between ET(P–M) and ET(H–S) in Giza governorate

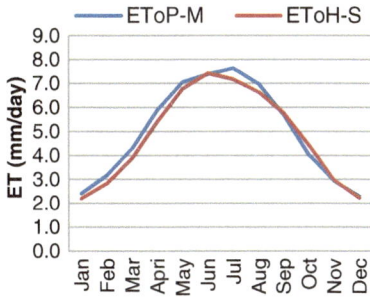

Fig. 1.16 Comparison between ET(P–M) and ET(H–S) in El-Fayoum governorate

Fig. 1.17 Comparison between ET(P–M) and ET(H–S) in Qena governorate

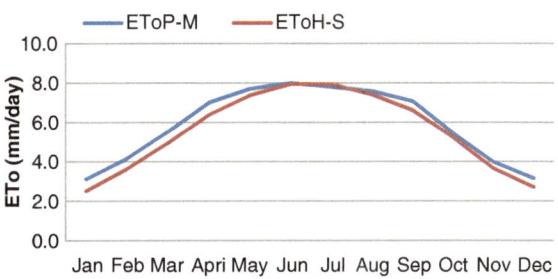

equation plotted as the independent variable. The intercept (a) and calibration slope (b) of the best fit regression line were used as regional calibration coefficients for each governorate. This methodology was developed by Shahidian et al. (2012) as follows:

$$ET(P - M) = a + b * ET(H-S). \tag{1.1}$$

An equation for each governorate was developed, where different (a) and (b) values were estimated. The quality of the fit between the two methodologies was presented in terms of the coefficient of determination (R^2), which is the ratio of the explained variance to the total variance and through calculation of root-mean-square error per observation (RMSE/obs), which gives the standard deviation of the model prediction error per observation (Jamieson et al. 1998). Regression equations,

Table 1.1 Prediction equations, coefficient of determination (R^2), and root-mean-square error per observation (RMSE/obs) for ET values in the studied governorates

Governorate	Prediction equation	R^2	RMSE/obs
Nile Delta			
Alexandria	ET(P–M) = −0.4252 + 1.2134*ET(H–S)	0.95	0.06
Demiatte	ET(P–M) = −0.2297 + 1.2714*ET(H–S)	0.98	0.05
Kafr El-Sheik	ET(P–M) = −0.1280 + 1.1690*ET(H–S)	0.98	0.05
El-Dakahlia	ET(P–M) = 0.0338 + 1.0745*ET(H–S)	0.98	0.04
El-Behira	ET(P–M) = 0.3432 + 1.1157*ET(H–S)	0.96	0.04
El-Gharbia	ET(P–M) = 0.2673 + 1.1019*ET(H–S)	0.97	0.04
El-Monofia	ET(P–M) = 0.0737 + 1.1640*ET(H–S)	0.98	0.05
El-Sharkia	ET(P–M) = 0.3484 + 1.0857*ET(H–S)	0.95	0.06
El-Kalubia	ET(P–M) = −0.1739 + 1.0498*ET(H–S)	0.99	0.03
Middle Egypt			
Giza	ET(P–M) = −0.1292 + 1.1050* ET(H–S)	0.99	0.04
Fayoum	ET(P–M) = 0.0702 + 1.0209* ET(H–S)	0.98	0.05
Beni Swief	ET(P–M) = 0.0015 + 1.0370* ET(H–S)	0.98	0.05
El-Minia	ET(P–M) = −0.0378 + 1.0033* ET(H–S)	0.98	0.04
Upper Egypt			
Assuit	ET(P–M) = 0.1352 + 1.1042* ET(H–S)	0.97	0.04
Suhag	ET(P–M) = −0.2569 + 1.0428* ET(H–S)	0.98	0.04
Qena	ET(P–M) = 0.7528 + 0.9270* ET(H–S)	0.99	0.03
Aswan	ET(P–M) = 0.2727 + 1.1500* ET(H–S)	0.97	0.03

coefficient of determination (R^2), and root-mean-square error per observation (RMSE/obs) are presented in Table (1.1).

The results from Table (1.1) showed that R^2 was between 0.95 and 0.99 in the Nile Delta. The variation in R^2 in Middle Egypt was between 0.98 and 0.99, whereas in Upper Egypt, the values of R^2 were between 0.97 and 0.99. Furthermore, RMSE/obs values were between 0.03 and 0.06 mm/day in Nile Delta. It was between 0.04 and 0.05 mm/day in Middle Egypt and between 0.03 and 0.04 mm/day in Upper Egypt. The presented equation for each governorate was used to predict monthly values of ET(P–M).

The calibration coefficients (*a* and *b*) with ET values calculated from Hargreaves–Samani equation could be used by other researchers to predict ET values similar to the one calculated by Penman–Monteith equation for each governorate in Egypt with high degree of accuracy because R^2 for each equation was close to 1 and RMSE/obs was close to zero (Table 1.1). The calibration coefficients (*a* and *b*) should be developed for each site to increase the accuracy of prediction of ET(P–M) values.

The predicted values of ET were compared with the estimated values of ET(P–M) and graphed together to ease comparison. Regarding to Alexandria and Demiatte governorates, the predicted values were close to calculated values in most of the months (Figs. 1.18 and 1.19).

Fig. 1.18 Comparison between ET(P–M) and predicted values of ET in Alexandria governorate

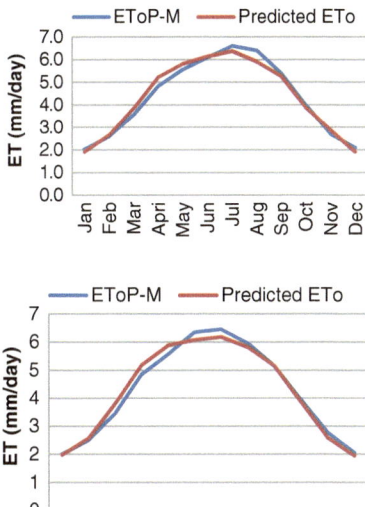

Fig. 1.19 Comparison between ET(P–M) and predicted values of ET in Demiatte governorate

Figures (1.20, 1.21, 1.22, 1.23, 1.24, 1.25, 1.26, 1.27, 1.28, 1.29, 1.30, 1.31, 1.32, 1.33, and 1.34) show that the deviation between calculated and predicted ET values was either low or not exist in the rest of governorates.

The results in the above graphs proved that the calibration coefficients were capable to account for the effect of relative humidity, wind speed, and potential sunshine hours, which not included in the H–S equation. ET values in the studied areas are presented in Table 1.2.

Fig. 1.20 Comparison between ET(P–M) and predicted values of ET in Kafr El-Sheik governorate

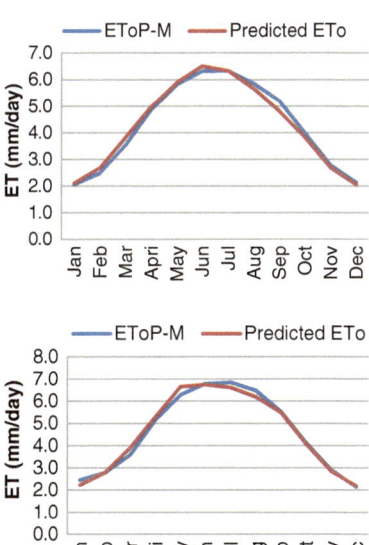

Fig. 1.21 Comparison between ET(P–M) and predicted values of ET in El-Dakahlia governorate

Fig. 1.22 Comparison
between ET(P–M) and
predicted values of ET in
El-Behira governorate

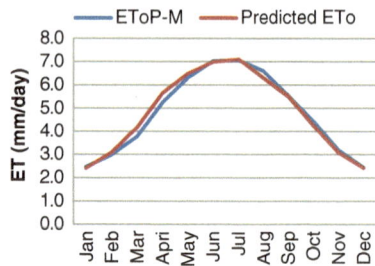

Fig. 1.23 Comparison
between ET(P–M) and
predicted values of ET in
El-Gharbia governorate

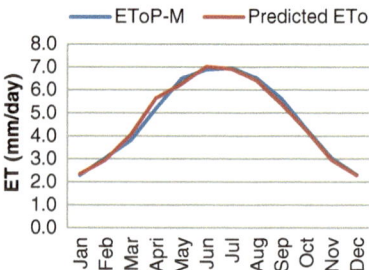

Fig. 1.24 Comparison
between ET(P–M) and
predicted values of ET in
Assuit governorate

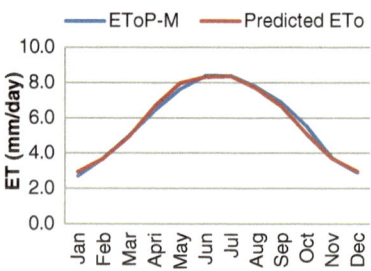

Fig. 1.25 Comparison
between ET(P–M) and
predicted values of ET in
Aswan governorate

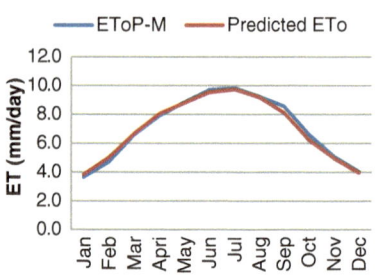

Fig. 1.26 Comparison between ET(P–M) and predicted values of ET in El-Monofia governorate

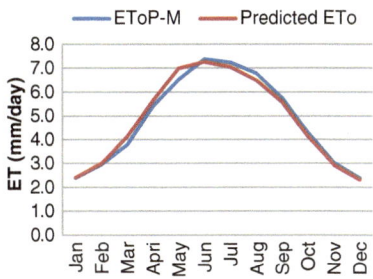

Fig. 1.27 Comparison between ET(P–M) and predicted values of ET in El-Sharkia governorate

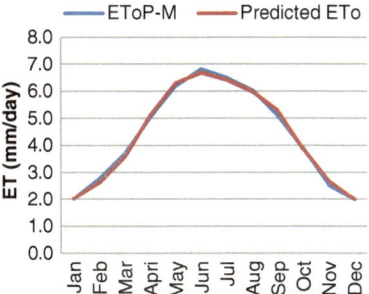

Fig. 1.28 Comparison between ET(P–M) and predicted values of ET in El-Kalubia governorate

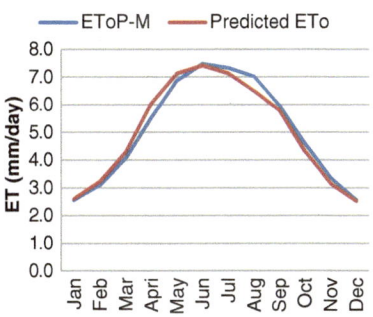

Fig. 1.29 Comparison between ET(P–M) and predicted values of ET in Beni Swief governorate

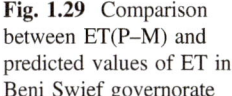

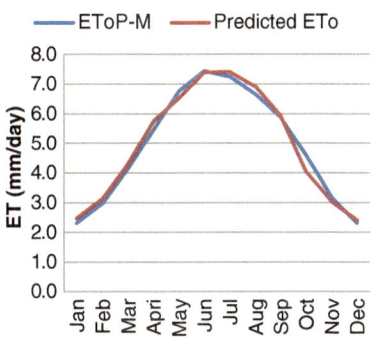

Fig. 1.30 Comparison between ET(P–M) and predicted values of ET in El-Minia governorate

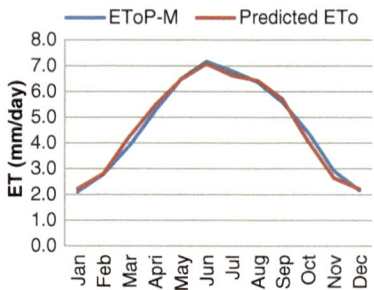

Fig. 1.31 Comparison between ET(P–M) and predicted values of ET in Suhag governorate

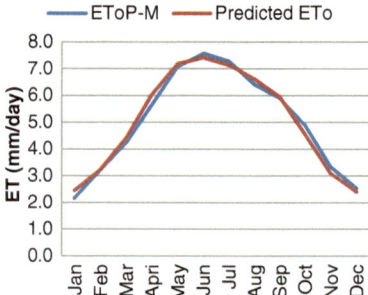

Fig. 1.32 Comparison between ET(P–M) and predicted values of ET in Giza governorate

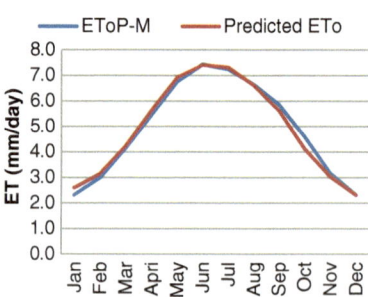

Fig. 1.33 Comparison between ET(P–M) and predicted values of ET in El-Fayoum governorate

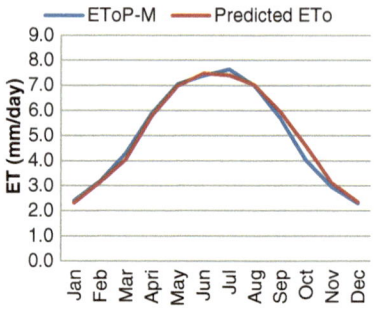

Fig. 1.34 Comparison between ET(P–M) and predicted values of ET in Qena governorate

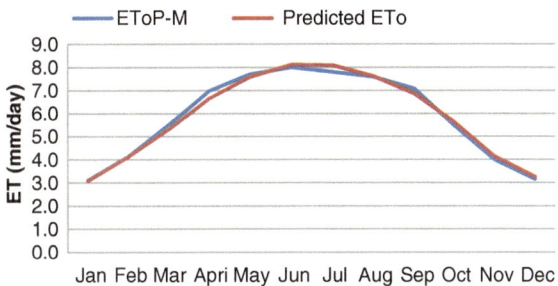

Table 1.2 Yearly average value of ET (mm/day) under current climate

Governorate	ET under current climate
Nile Delta	
Alexandria	4.32
Demiatte	4.25
Kafr El-Sheik	4.28
El-Dakahlia	4.59
El-Behira	4.79
El-Gharbia	4.71
El-Monofia	4.83
El-Sharkia	4.38
El-Kalubia	5.01
Middle Egypt	
Giza	4.91
Fayoum	5.01
Beni Swief	4.94
El-Minia	4.66
Upper Egypt	
Assuit	5.76
Suhag	5.04
Qena	5.87
Aswan	7.01

Khalil (2013) indicated that under current climate, Aswan governorate has the highest value of ET, in comparison with all other governorates, and Demiatte has the lowest value of ET, which is similar to what is presented in Table (1.2).

Similar procedure could be done using ET(P–M) values obtained from FAO AQUASTAT website: http://www.fao.org/nr/water/aquastat/gis/index3.stm. These values can be obtained for any location on Earth. The obtained values are normal weather parameters (average of 30 years from 1961 to 1990), in addition to ET values calculated using P–M equation.

Furthermore, the above methodology could solve a large problem that faces researchers and extension workers in irrigation scheduling in Egypt and in other developing countries. The availability of a number of meteorological stations, to measure weather parameters, is limited, and reliability of the measured data could be an obstacle. There are also concerns about the accuracy of the observed meteorological parameters (Droogers and Allen 2002), since the actual instruments, specifically pyranometers (solar radiation) and hygrometers (relative humidity), are often subject to stability errors, where it is common to see a drift as high as 10 % in pyranometers (Samani 2000). Sepaskhah and Razzaghi (2009) have observed that hygrometers lose about 1 % in accuracy per installed month. Thus, they recommend the use of ET equations that require fewer variables. Hargreaves and Allen (2003) concluded that the differences in ET values, calculated by the different methods, are minor when compared with the uncertainties in estimating actual crop evapotranspiration from measured weather data. Additionally, these equations can be more easily used in adaptive or smart irrigation controllers that adjust the application depth according to the daily ET demand (Shahidian et al. 2009).

Evapotranspiration Under Climate Change

The agricultural system in Egypt is vulnerable to climate change due to its limited water resources and strong dependence on irrigation for crop production. Exploring the impacts of climate change on crop evapotranspiration is important for water management and agricultural sustainability. Climate change and its syndrome, i.e., higher temperature, will increase ET and that will affect the hydrological system and water resources (Shahid 2011). In Egypt, temperature rise by 1 °C may increase ET rate by about 4–5 % (Eid 2001). Furthermore, Khalil (2013) indicated that ET values will increase under climate change compared to current climate. Thus, quantifying the changes in ET due to climate change is very important for management of water resources.

Climate Change Model

Research program on Climate Change Agriculture and Food Security (CCAFS) is one of CIGAR programs that implement a uniquely innovative and transformative research program that addresses agriculture in the context of climate variability, climate change, and uncertainty about future climate conditions. The details of the methodology are presented in Jones et al. (2009). The link to this web site is the following: http://www.ccafs.cgiar.org/marksimgcm#.Ujh1gj-GfMY. The web site is composed of seven global climate change models. For each model, three climate change scenarios (A1B, A2, and B1) can be downloaded.

The climate model ECHAM5 (Roeckner et al. 2003) is one of them and was used in this analysis. The model is Atmospheric Oceanic General Circulation model. It has been developed from the ECMWF operational forecast model cycle

36 (1989) (therefore, the first part of its name: EC) and a comprehensive parameterization package developed at Hamburg (therefore, the abbreviation HAM). The part describing the dynamics of ECHAM is based on the ECMWF documentation, which has been modified to describe the newly implemented features and the changes necessary for climate experiments. Since the release of the previous version, ECHAM4, the whole source code, has been extensively redesigned in the major infrastructure and transferred to FORTRAN 95, ECHAM is now fully portable and runs on all major high-performance platforms. The restart mechanism is implemented on top of net CDF and because of that it absolutely independent on the underlying architecture. The resolution of the model is 1.9 × 1.9°.

Climate Change Scenario

ECHAM5 model was used to develop A1B climate change scenario for each weather station in each governorate. IPCC (2007) describes the A1 storyline and scenario family as a future world of very rapid economic growth, global population that peaks in mid-century and declines thereafter, and the rapid introduction of new and more efficient technologies. Major underlying themes are convergence among regions, capacity building, and increased cultural and social interactions, with a substantial reduction in regional differences in per capita income. One of its family is A1B, where its technological balance is across all sources (balanced is defined as not relying too heavily on one particular energy source, on the assumption that similar improvement rates apply to all energy supply and end-use technologies).

The downloaded scenario was for the years 2020, 2030, and 2040 and composed of maximum and minimum temperature, rain, and solar radiation. These weather parameters were not enough to calculate ET with P–M equation. However, they are enough to calculate ET using H–S equation. The developed equations were used to calculate ET values under A1B climate change scenario in 2020, 2030, and 2040.

Calculation of ET Under A1B Climate Change Scenario

ET values under A1B climate change scenario in 2020, 2030, and 2040 were calculated for each governorate, and yearly average value was calculated. Figures (1.35, 1.36, 1.37, 1.38, 1.39, 1.40, 1.41, 1.42, 1.43, 1.44, 1.45, 1.46, 1.47, 1.48, 1.49, 1.50 and 1.51) presented comparison between ET values under current climate and under A1B climate change scenario 2020, 2030, and 2040 in each governorate.

Fig. 1.35 ET values under current climate and A1B climate change scenario in Alexandria governorate

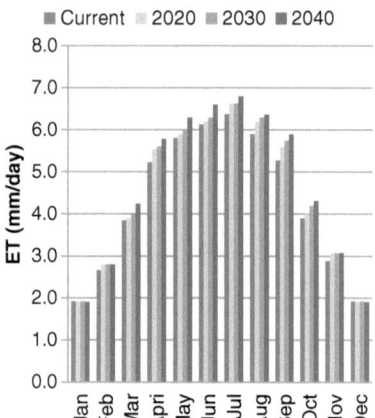

Fig. 1.36 ET values under current climate and A1B climate change scenario in Demiatte governorate

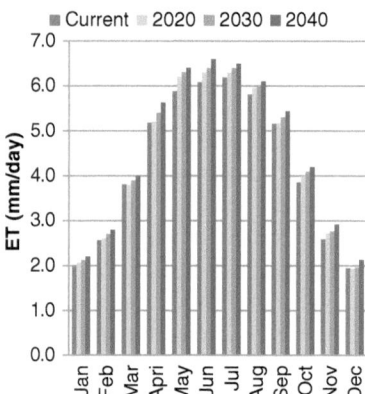

Fig. 1.37 ET values under current climate and A1B climate change scenario in Kafr El-Sheik governorate

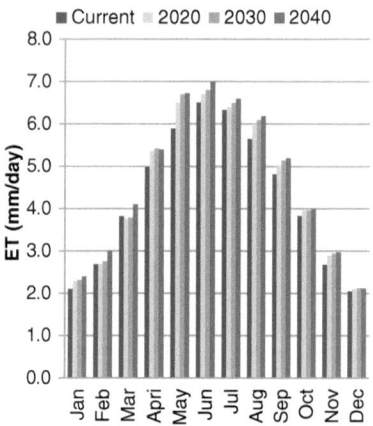

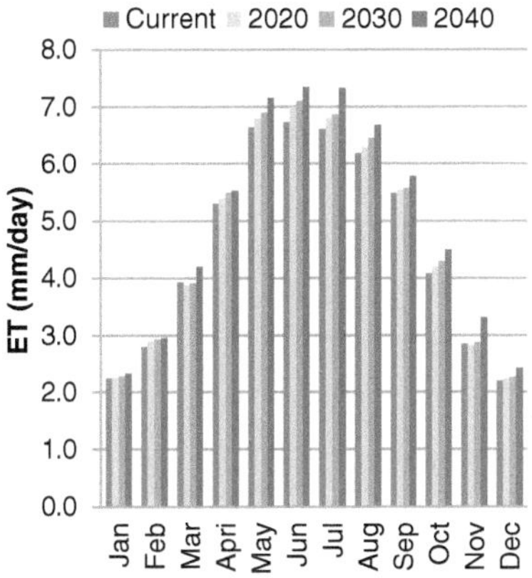

Fig. 1.38 ET values under
current climate and A1B
climate change scenario in
El-Dakahlia governorate

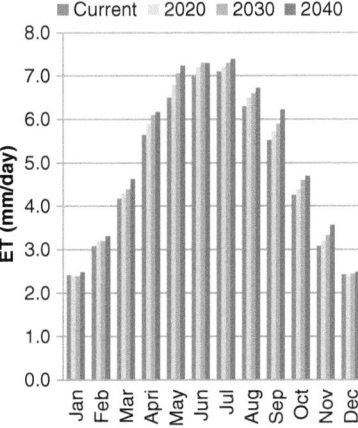

Fig. 1.39 ET values under
current climate and A1B
climate change scenario in
El-Behira governorate

Fig. 1.40 ET values under
current climate and A1B
climate change scenario in
El-Gharbia governorate

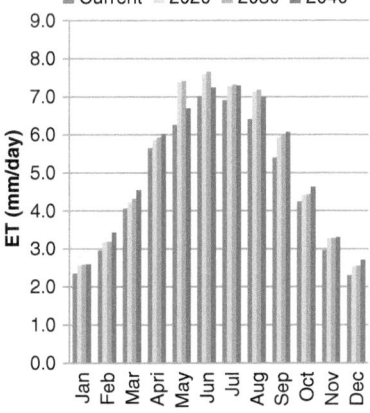

Fig. 1.41 ET values under current climate and A1B climate change scenario in Assuit governorate

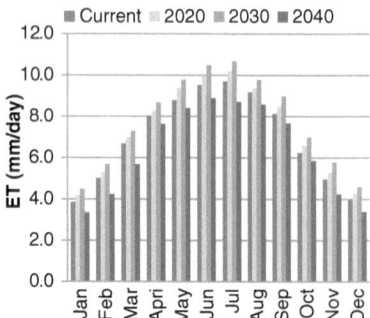

Fig. 1.42 ET values under current climate and A1B climate change scenario in Aswan governorate

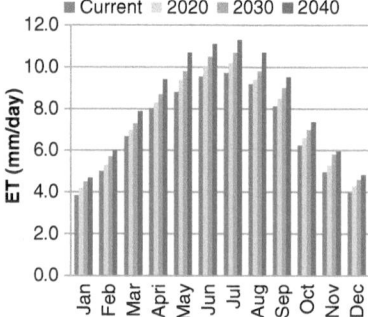

Fig. 1.43 ET values under current climate and A1B climate change scenario in El-Monofia governorate

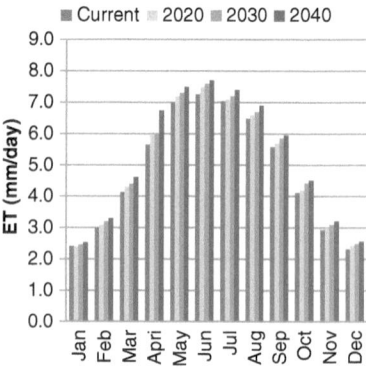

These graphs showed that the value of monthly ET under climate change increases gradually starting from January until April and then the increase become higher during the summer months. In the fall months until December, the decrease became lower. This trend was found in all studied governorates.

The above graphs implied that under A1B climate change scenario, the value of ET will increase, and consequently water requirements for crops are expected to

Fig. 1.44 ET values under current climate and A1B climate change scenario in El-Sharkia governorate

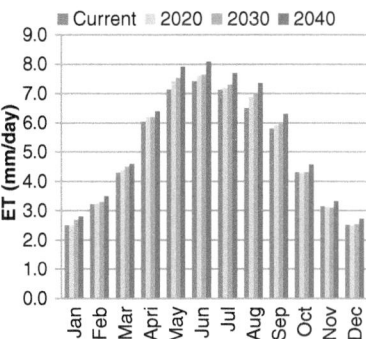

Fig. 1.45 ET values under current climate and A1B climate change scenario in El-Kalubia governorate

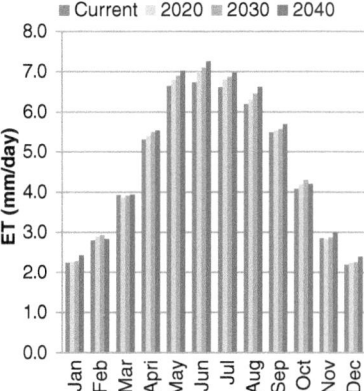

Fig. 1.46 ET values under current climate and A1B climate change scenario in Beni Swief governorate

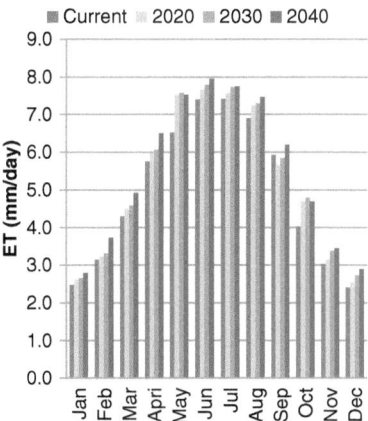

Fig. 1.47 ET values under current climate and A1B climate change scenario in El-Minia governorate

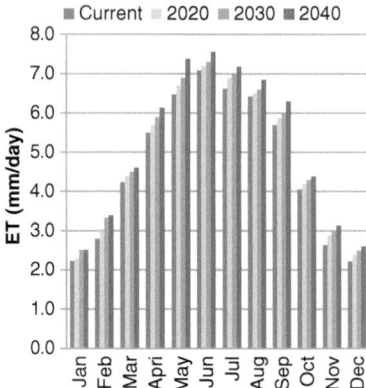

Fig. 1.48 ET values under current climate and A1B climate change scenario in Suhag governorate

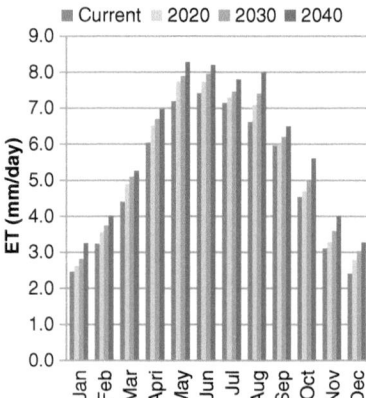

Fig. 1.49 ET values under current climate and A1B climate change scenario in El-Giza governorate

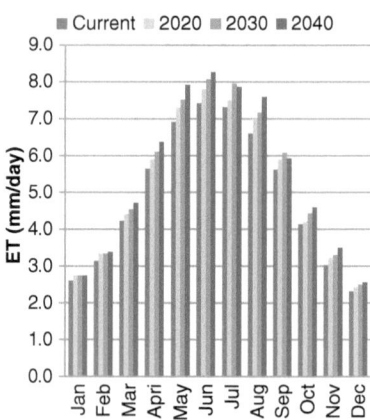

Fig. 1.50 ET values under current climate and A1B climate change scenario in El-Fayoum governorate

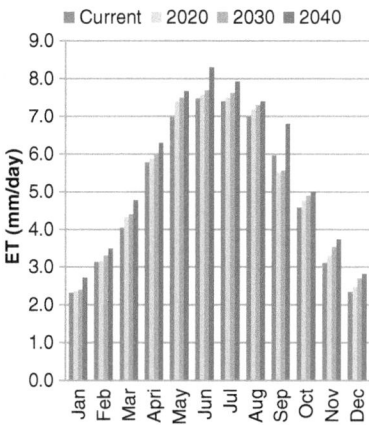

Fig. 1.51 ET values under current climate and A1B climate change scenario in Qena governorate

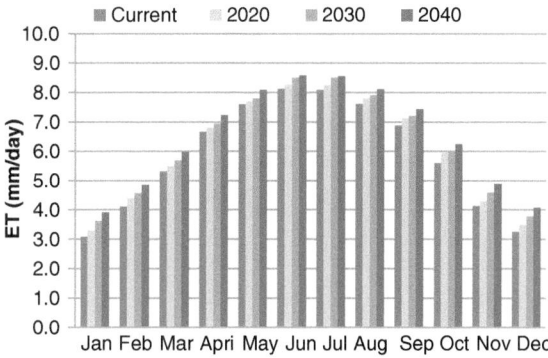

increase in all governorates with different values. Table (1.3) presents the percentage of increase in ET annual values in all governorates. The average ET values were 7, 9, and 13 % in 2020, 2030, and 2040, respectively. Snyder et al. (2011) concluded that the impact of global warming on ET will likely be less in locations with higher wind speeds. The northern five governorates in the Nile Delta are located on the Mediterranean Sea and characterized by high wind speeds between 4.3 and 4.9 m s^{-1}. Furthermore, the percentage of increase in ET under A1B climate change scenario was the lowest in the three tested future years in these five governorates.

Table 1.3 Percentage of increase in ET (mm/day) under A1B climate change scenario in 2020, 2030, and 2040

Governorate	ET in 2020	ET in 2030	ET in 2040
Nile Delta			
Alexandria	1	3	4
Demiatte	1	2	2
Kafr El-Sheik	2	2	2
El-Dakahlia	1	1	2
El-Behira	1	2	2
El-Gharbia	7	10	19
El-Monofia	5	10	19
El-Sharkia	7	14	17
El-Kalubia	7	11	19
Middle Egypt			
Giza	8	15	16
Fayoum	8	12	16
Beni Swief	10	13	18
El-Minia	10	15	19
Upper Egypt			
Assuit	11	12	17
Suhag	12	10	18
Qena	11	14	19
Aswan	12	14	19
Average	7	9	13

Conclusion

Quantification of the impact of climate change on ET is very important for policy makers when developing future plans. This requires an accurate equation to calculate ET values. With only monthly maximum and minimum temperature measurements and solar radiation available, monthly ET can be calculated by H–S equation and then regressed on ET value previously calculated from P–M equation to develop calibration coefficients for each site. The above results showed that this method was accurate and the predicted ET values were close to the calculated ET values by P–M equation.

References

Allen RG, Jensen ME, Wright JL, Burman RD (1989) Operational estimate of reference evapotranspiration. Agron J 81:650–662

Allen RG, Pereira LS, Raes D, Smith M (1998) Crop evapotranspiration: guideline for computing crop water requirements. FAO N°56

Blaney HF, Criddle WD (1950) Determining water requirements in irrigated areas from climatological and irrigation data. USDA/SCS, SCS-TP 96

Droogers P, Allen RG (2002) Estimating reference evapotranspiration under inaccurate data conditions. Irrig Drain Syst 16(1):33–45

Eid HM (2001) Climate change studies on Egyptian agriculture. Soils, Water and Environment Research Institute. Agricultural Research Center, Egypt

Gardner FP, Pearce RB, Mitchell RL (1985) Physiology of crop plants. IOWA State University Press, Ames

Hargreaves GH, Allen RG (2003) History and evaluation of Hargreaves evapotranspiration equation. J Irrig Drain Eng 129(1):53–63

Hargreaves GH, Samani ZA (1982) Estimating potential evapotranspiration. J Irrig Drain Div 108 (3):225–230

Hargreaves GH, Samani ZA (1985) Reference crop evapotranspiration from temperature. Trans ASAE 1(2):96–99

IPCC Intergovernmental Panel on Climate Change (2007) Intergovernmental panel on climate change fourth assessment report: climate change 2007. Synthesis report. World Meteorological Organization, Geneva

Jamieson PD, Porter JR, Goudriaan J, Ritchie JT, van Keulen H, Stol W (1998) A comparison of the models AFRCWHEAT2, CERES-Wheat, Sirius, SUCROS2 and SWHEAT with measurements from wheat grown under drought. Field Crops Res 55:23–44

Jones PG, Thornton PK, Heinke J (2009) Generating characteristic daily weather data using downscaled climate model data from the IPCC's fourth assessment: project report

Khalil AA (2013) Effect of climate change on evapotranspiration in Egypt. Researcher 5(1):7–12

Khalil F, Ouda SA, Osman N, Khamis E (2011) Determination of agro-climatic zones in Egypt using a robust statistical procedure. In: 15th international conference on water technology, Alexandria, 30 May–2 June

Monteith JL (1965) Evaporation and environment. In: Fogg GE (ed) Symposium of the Society for Experimental Biology: the state and movement of water in living organisms, vol 19. Academic Press Inc, NY, pp 205–234

Roeckner E, Bäuml G, Bonaventura L, Brokopf R, Esch M, Giorgetta M, Hagemann S, Kirchner I, Kornblueh L, Manzini E, Rhodin A, Schlese U, Schulzweida U, Tompkins A (2003) The atmospheric general circulation model ECHAM5. Part I: model description. MPI Report 349, Max Planck Institute for Meteorology, Hamburg, Germany, 127 pp

Samani Z (2000) Estimating solar radiation and evapotranspiration using minimum climatological data. J Irrig Drain Eng 126(4):265–267

Sepaskhah AR, Razzaghi FH (2009) Evaluation of the adjusted Thornthwaite and Hargreaves-Samani methods for estimation of daily evapotranspiration in a semiarid region of Iran. Arch Agron Soil Sci 55(1):51–56

Shahid S (2011) Impacts of climate change on irrigation water demand in Northwestern Bangladesh. Clim Change 105(3–4):433–453

Shahidian S, Serralheiro R, Teixeira JL, Santos FL, Oliveira MR, Costa J, Toureiro C, Haie N, Machado R (2009) Drip irrigation using a PLC based adaptive irrigation system WSEAS transactions on environment and development, vol 2

Shahidian S, Serralheiro R, Serrano J, Teixeira J, Haie N, Francisco S (2012) Hargreaves and other reduced-set methods for calculating evapotranspiration. In: Irmak A (ed) Evapotranspiration—remote sensing and modeling. ISBN: 978-953-307-808-3, InTech. http://www.intechopen.com/books/evapotranspiration-remote-sensing-and-modeling/hargreaves-and-otherreduced-set-methods-for-calculating-evapotranspiration

Snyder RL, Orang M, Bali K, Eching S (2004) Basic irrigation scheduling (BIS). http://www.waterplan.water.ca.gov/landwateruse/wateruse/Ag/CUP/Californi/Climate_Data_010804.xls

Snyder RL, Moratiel R, Zhenwei S, Swelam A, Jomaa I, Shapland T (2011) Evapotranspiration response to climate change. Acta Hortic (ISHS) 922:91–98. http://www.actahort.org/books/922/922_11.htm

Chapter 2
Water Requirements for Major Crops

Samiha Ouda, Khaled Abd El-Latif and Fouad Khalil

Abstract The objective of this chapter was to calculate water requirements for four crops: wheat, maize, rice and sugarcane grown in 17 governorates in Egypt under current climate and under the A1B climate change scenario in 2040. The BISm model was used to calculate crop coefficient water depletion from root zone. Water requirements under A1B climate change were calculated using the model. The results indicated that water requirements for wheat will increase by 2–19% depending on governorate location. The effect of climate change was more pronounced on maize as a summer crop, where the applied irrigation amount is expected to increase in all governorates under climate change in 2040 by 10–19%. Regarding rice, water requirements were increased by 10–14%. With respect to sugarcane, which is a unique case because of its long growing season (365 days), its water requirements under climate change conditions increased by 11–19%.

Keywords Wheat · Maize · Rice · Sugarcane

The term crop water requirement is defined as the amount of water required to compensate the evapotranspiration loss from the cropped field (USDA Soil Conservation Service 1993). The ICID-CIID (2000) describes it as the total water needed for evapotranspiration, from planting to harvest for a given crop in a specific climate, when adequate soil water is maintained by rainfall and/or irrigation so that it does not limit plant growth and crop yield. Although the values for crop evapotranspiration and crop water requirement are identical, crop water requirement refers to the amount of water that needs to be supplied, while crop evapotranspiration (ETc) refers to the amount of water that is lost through evapotranspiration (Allen et al. 1998). Furthermore, in estimating crop water requirements, efficiency of the irrigation system should be taken into account.

ETc accounts for variations in weather and offers a measure of the "evaporative demand" of the atmosphere, whereas crop coefficient (kc) accounts for the difference between ETc and ET (Snyder et al. 2004). The kc takes into account the

S. Ouda (✉) · K.A. El-Latif · F. Khalil
Water Requirements and Field Irrigation Research Department, Soils,
Water and Environment Research Institute, Agricultural Research Center, Giza, Egypt

© The Author(s) 2016   25
S. Ouda, *Major Crops and Water Scarcity in Egypt*,
SpringerBriefs in Water Science and Technology,
DOI 10.1007/978-3-319-21771-0_2

relationship between atmosphere, crop physiology, and agricultural practices (Lascano 2000). Therefore, sowing date, which reflects the weather of a certain site, could affect growth pattern of a crop and consequently affects the period of growth stages, the value of kc, and growth period. As a crop canopy develops, the ratio of transpiration (T) to ET increases until most of the ET comes from T and evapotranspiration (E) is a minor component. This occurs because light interception by the foliage increases until most light is intercepted before it reaches the soil (Snyder et al. 2004). Therefore, crop coefficients for field crops generally increase until the canopy ground cover reaches about 75 % and the light interception is near 80 %. The accurate calculation of crop kc for each growth stage is an important component for accurate calculation of water requirements (Shideed et al. 1995).

Irrigation would help maintain optimal soil moisture during the growing period, thereby ensuring a more stable and higher agricultural production. Irrigation and its planning are demanding tasks, which involve a multidisciplinary approach to define and calculate all the relevant factors and parameters. The evidence of future warming is another reason for paying much attention to the efficiency of water use by plants (Katerji and Rana 2008).

The objective of this chapter was to calculate water requirements for four crops: wheat, maize, rice, and sugarcane grown in 17 governorates in Egypt under current climate and under A1B climate change scenario in 2040.

BISm Model

The required irrigation water need to be applied to the studied crops was estimated using BISm model (Snyder et al. 2004). The model requires planting and harvesting dates as input. The model calculates crop coefficient (kc). The model also account for water depletion from root zone. Therefore, it requires to input total water holding capacity and available water in the soil. These values were obtained from previous research done in the Water Requirements and Field irrigation Research Department, Soils, Water and Environment Research Institute, Agricultural Research Center, Egypt (Table 2.1).

Planting and harvesting dates under current climate for the studied crops in each of the 17 governorates were obtained from bulletins published by Agricultural Research Center. Wheat and maize are cultivated in all governorates. The season length for wheat, as winter crop, is 155 days. Maize season length is 110 days. Planting and harvesting dates are included in Table 2.2.

Season length for rice is 140 days and cultivated in the all Nile Delta governorates, except El-Monofia governorate. It is cultivated on 15th of May and harvested on 30th of September. Sugarcane is an annual crop planted for sugar industry in 4 governorates, i.e., El-Minia, Suhag, Qena, and Aswan. Spring sugarcane is cultivated on 15th of February and harvested on 14th of February in the following year.

Table 2.1 Soil water holding capacity and available water prevailed in each governorate

Governorate	Water holding capacity (m/m)	Available water (m/m)
Nile Delta		
Alexandria	0.373	0.206
Demiatte	0.376	0.222
Kafr El-Sheik	0.405	0.170
El-Dakahlia	0.395	0.196
El-Behira	0.408	0.230
El-Gharbia	0.380	0.220
El-Monofia	0.418	0.232
El-Sharkia	0.420	0.210
El-Kalubia	0.400	0.218
Middle Egypt		
Giza	0.363	0.209
Fayoum	0.426	0.194
Beni Swief	0.429	0.245
El-Minia	0.435	0.239
Upper Egypt		
Assuit	0.438	0.235
Suhag	0.446	0.244
Qena	0.454	0.293
Aswan	0.447	0.257

Table 2.2 Planting and harvesting dates for wheat and maize under current climate conditions

Governorate	Wheat		Maize	
	Planting date	Harvest date	Planting date	Harvest date
Nile Delta and Middle Egypt	15-Nov	18-Apr	15-May	01-Sep
Upper Egypt	01-Nov	01-Apr	01-May	18-Aug

A1B climate change scenario was developed using ECHAM5 climate model (Roeckner et al. 2003) in 2040 for each 17 governorates in the Nile Delta and Valley. ECHAM5 model is an Atmospheric Oceanic General Circulation model with low resolution, i.e., $1.9 \times 1.9°$. Water requirements under climate change were calculated using BISm model (Snyder et al. 2004). The same planting date was assumed under climate change, and harvesting date was assumed to be one week earlier for all crops, except sugarcane. Previous research on the effect of climate change on season length indicated that it could be reduced by 7–12 days (Khalil et al. 2009; Ouda et al. 2009). Water requirements for selected crops were calculated under surface irrigation, where application efficiency is 60 % (Abou Zeid 2002). The calculations were done under current climate and under climate change scenario A1B in 2040.

Water Requirements for Wheat

Table 2.3 presents the applied amount of irrigation water for wheat in 2012/13 growing season and the value of water requirements in 2040. The lowest percentage of increase between applied and required water (PI %) was found in the governorates that have seashore on the Mediterranean Sea. In these governorates, the difference between ET under current climate and under climate change during wheat growing season was low. Starting from El-Gharbia governorate, the difference between ET under current climate and under climate change became higher. Furthermore, the ET value for March in El-Gharbia governorate and the rest of governorates was relatively higher, compared to the value of February. Going south of Egypt, PI % became high, i.e., 17–19 %.

Ouda et al. (2010) and Ibrahim et al. (2012) calculated water requirements for wheat grown in El-Behira governorate under climate change in 2030. They both concluded that it would increase by an average of 3–4 % under A2 and B2 climate change scenarios developed by Hadley model. In Demiatte governorate, the water requirement for wheat is expected to increase under climate change condition by

Table 2.3 Applied irrigation water for wheat under current climate, water requirements under A1B in 2040, and percentage of increase (PI %)

Governorate	Applied irrigation under current climate (mm)	Water requirements under climate change (mm)	PI(%)
Nile Delta			
Alexandria	462	478	3
Demiatte	495	505	2
Kafr El-Sheik	510	521	2
El-Dakahlia	519	528	2
El-Behira	501	513	2
El-Gharbia	497	585	18
El-Monofia	521	618	19
El-Sharkia	560	658	18
El-Kalubia	580	689	19
Middle Egypt			
Giza	556	664	19
Fayoum	516	594	15
Beni Swief	544	644	18
El-Minia	531	631	19
Upper Egypt			
Assuit	625	728	16
Suhag	546	645	18
Qena	654	776	19
Aswan	803	953	19

3 % (Noreldin et al. 2013). Similar results were obtained by Khalil et al. (2009), where water requirements for wheat grown in Giza were increased by 10 % under climate change conditions in 2030.

Water Requirements for Maize

The effect of climate change was more pronounced on maize as a summer crop. The applied irrigation amount for maize is expected to increase in all governorates under climate change in 2040. The highest percentage of increase in the applied irrigation water in the Nile Delta existed in El-Behira, El-Monofia, and El-Kalubia, i.e., 15 %. Similar results were obtained by Ouda et al. (2012) and Abdrabbo et al. (2013) under El-Behira governorate. A value of 16 % increase in the water requirements under climate change conditions existed in all Middle Egypt and Upper Egypt governorates, except for Aswan, where the PI % value was 19 % (Table 2.4). This could be attributed to the difference between ET under current climate and under climate change conditions which was high in all months of the year at Aswan. Furthermore, Aswan is characterized by having hottest summer days in Egypt.

Table 2.4 Applied irrigation water for maize under current climate, water requirements under A1B in 2040, and percentage of increase (PI %)

Governorate	Applied irrigation under current climate (mm)	Water requirements under climate change (mm)	PI (%)
Nile Delta			
Alexandria	772	859	11
Demiatte	735	812	10
Kafr El-Sheik	780	865	11
El-Dakahlia	703	797	13
El-Behira	728	838	15
El-Gharbia	710	809	14
El-Monofia	734	846	15
El-Sharkia	764	869	14
El-Kalubia	802	925	15
Middle Egypt			
Giza	871	1013	16
Fayoum	879	1018	16
Beni Swief	881	1021	16
El-Minia	896	1040	16
Upper Egypt			
Assuit	895	1039	16
Suhag	884	1023	16
Qena	890	1035	16
Aswan	1090	1295	19

Water Requirements for Rice

The value of PI (%) between applied irrigation water for rice under current climate and water requirements under climate change was high (Table 2.5). The value was between 10 and 14 %, with the lowest value in Alexandria and Demiatte, i.e., 10 % and the highest value existed in El-Kalubia governorate.

Water Requirements for Sugarcane

Sugarcane is a unique case because its growing season is 365 days and it is cultivated in four governorates only in south Egypt. The value of PI (%) between applied water under current climate and water requirements under climate change conditions was between 11 and 19 %. The highest value was in Aswan, i.e., 19 % (Table 2.6).

Table 2.5 Applied irrigation water for rice under current climate, water requirements under A1B in 2040, and percentage of increase (PI %)

Governorate	Applied irrigation under current climate (mm)	Applied irrigation under climate change (mm)	PI (%)
Alexandria	1045	1146	10
Demiatte	1018	1124	10
Kafr El-Sheik	1031	1141	11
El-Dakahlia	1116	1234	11
El-Behira	1162	1290	11
El-Gharbia	1104	1236	12
El-Sharkia	1145	1279	12
El-Kalubia	1113	1267	14

Table 2.6 Applied irrigation water for sugarcane under current climate, under A1B climate change scenario, and percentage of increase (PI)

Governorate	Applied irrigation under current climate (mm)	Applied irrigation under climate change (mm)	PI (%)
El-Minia	3140	3488	11
Suhag	3333	3758	13
Qena	3824	4246	11
Aswan	4680	5553	19

Conclusion

Our results indicated that climate change condition in 2040 is expected to increase ET values for all governorates in Egypt. Consequently, water requirements will also increase. The lowest increase in water requirements was at the Nile Delta, compared to Middle Egypt and Upper Egypt.

In semiarid regions, where Egypt is located, more pressure will be put on water resources distribution between economic sectors under climate change, especially agriculture. Reduction in the amount of allocated water for irrigation, increase in water requirements for crops, and yield reduction under climate change conditions will worsen food security situation in Egypt. Since surface irrigation is prevailing in Egypt with low application efficiency, it is important to test options to increase it. These options, if implemented, will save on the applied irrigation water, which can be used to irrigate new lands under climate change conditions, which will contribute in maintaining the current production. Thus, it is very important to revise and fix the production system for cultivated crops, in terms of the used cultivars, fertilizer, and irrigation application.

References

Abdrabbo M, Ouda S, Noreldin T (2013) Modeling the irrigation schedule on wheat under climate change. Nat Sci 115:10–18

Abou Zeid K (2002) Egypt and the world water goals. Egypt statement in the world summit for sustainable development and beyond, Johannesburg

Allen RG, Pereira LS, Raes D, Smith M (1998) Crop evapotranspiration: guideline for computing crop water requirements. FAO N°56

Ibrahim M, Ouda S, Taha A, El Afandi G, Eid SM (2012) Water management for wheat grown in sandy soil under climate change conditions. J Soil Sci Plant Nutr 122:195–210

ICID-CIID (2000) Multilingual technical dictionary on irrigation and drainage—CD version September 2000. In: International commission on irrigation and drainage, New Dehli

Katerji N, Rana G (2008) Crop evapotranspiration measurement and estimation in the Mediterranean region, ISBN 978 8 89015 2412. INRA-CRA, Bari

Khalil FA, Farag H, El Afandi G, Ouda SA (2009) Vulnerability and adaptation of wheat to climate change in Middle Egypt. In: 13th conference on water technology. Hurghada, Egypt, pp 71–88

Lascano RJ (2000) A general system to measure and calculate daily crop water use. Agron J 92:821–832

Noreldin T, Ouda S, Abou Elenein R (2013) Development of management practices to address wheat vulnerably to climate change in north Delta. In: 11th international conference on development of dryland. Beigin, China

Ouda S, Khalil F, Yousef H (2009) Using adaptation strategies to increase water use efficiency for maize under climate change conditions. In: 13th international conference on water technology. Hurghada, Egypt

Ouda S, Sayed M, El Afandi G, Khalil F (2010) Developing an adaptation strategy to reduce climate change risks on wheat grown in sandy soil in Egypt. In: 10th international conference on development of dry lands. Cairo, Egypt

Ouda S, Abdrabbo M, Noreldin T (2012) Effect of changing sowing dates and irrigation scheduling on maize yield grown under climate change conditions. In: 4th International Conference for Field Irrigation and Agricultural Meteorology. Ameria, Egypt

Roeckner E, Bäuml G, Bonaventura L, Brokopf R, Esch M, Giorgetta M, Hagemann S, Kirchner I, Kornblueh L, Manzini E, Rhodin A, Schlese U, Schulzweida U, Tompkins A (2003) The atmospheric general circulation model ECHAM5. Part I: model description. MPI Report 349, Max Planck Institute for Meteorology, Hamburg, Germany, 127 pp

Shideed K, Oweis T, Gabr M, Osman M (1995) Assessing on-farm water use efficiency: a new approach, ed. ICARDA/ESCWA, Aleppo, Syria, 86 pp

Snyder RL, Orang M, Bali K, Eching S (2004) Basic Irrigation Scheduling BISm. http://www. waterplan.water.ca.gov/landwateruse/wateruse/Ag/CUP/Californi/Climate_Data_010804.xls

USDA Soil Conservation Service (1993) Irrigation water requirements. National Engineering Handbook NEH, Part 623, chapter 2, National Technical Information Service

Chapter 3
Significance of Reduction of Applied Irrigation Water to Wheat Crop

Samiha Ouda and Abd El-Hafeez Zohry

Abstract In this chapter, we investigated options to reduce the applied water for wheat, such as cultivation on raised beds and using sprinkler system for irrigation. We also investigated the effect of these two options on wheat national production under climate change in 2040. The effect of relay intercropping cotton on wheat on national production was also explored under current climate and under climate change. We also examined the effect of these options on water productivity under current climate and in 2040. The results indicated that under current situation of wheat production, which is grown under surface irrigation, production represents only 62 % of total consumption. Cultivating wheat on raised beds or using sprinkler system for irrigation increased total production compared to its counterpart under surface irrigation, which resulted in reduction of the applied water, increase in productivity and using the saved water in new land irrigation. Under climate change, wheat production will be reduced under surface irrigation. However, cultivation on raised beds or using sprinkler system for irrigation under climate change compensated a part of these losses. The results also indicated that the potential available water to irrigate new lands after changing water management practice can be reduced under climate change, compared to its values under current climate. As a result, the potential new cultivated areas will also be reduced. In relay intercropping cotton on wheat, the total cultivated area of wheat will consist of wheat area, cotton area and the added area as a result of water availability from raised bed cultivation WHICH will increase wheat total production under current and in 2040. The results also indicate that water productivity was the lowest under surface irrigation and was the highest when sprinkler system was used under both current and climate change.

Keywords Surface irrigation · Raised beds cultivation · Sprinkler system · Relay intercropping cotton on wheat · Water productivity

S. Ouda (✉)
Water Requirement and Field Irrigation Department, Soils, Water and Environment Research Institute, Agricultural Research Center, Giza, Egypt

A.E.-H. Zohry
Crop Intensification Department, Field Crops Research Institute, Agricultural Research Center, Giza, Egypt

S. Ouda, *Major Crops and Water Scarcity in Egypt*,
SpringerBriefs in Water Science and Technology,
DOI 10.1007/978-3-319-21771-0_3

Wheat is Egypt's major staple crop supplies more than one-third of the daily caloric intake and 45 % of total daily protein consumption of consumers. Wheat is planted as a winter crop and occupies about 33 % of the total winter crop area in Egypt. In 2012, the cultivated area of wheat in the Nile Delta and Valley was 1,109,570 ha, and the average productivity was 5.86 ton/ha. There is a gap between production and consumption of wheat in Egypt estimated by 40 %.

Climate change will alter the normal growing conditions for wheat, which will result in abiotic stress, such as heat and water stresses. Exposing wheat plants to high moisture stress depressed seasonal consumptive use and grain yield (Bukhat 2005). During vegetative growth, phyllochron decreases in wheat under water stress (McMaster 1997) and leaves become smaller, which could reduce leaf area index (Gupta et al. 2001) and number of reproductive tillers, in addition to limit their contribution to grain yield (Dencic et al. 2000). Furthermore, wheat is very sensitive to high temperature (Slafer and Satorre 1999). Wheat experiences heat stress to varying degrees at different phenological stages, but heat stress during the reproductive phase is more harmful than during the vegetative phase due to the direct effect on grain number and dry weight (Wollenweber et al. 2003).

Understanding the potential impacts of climate change is very important in developing both adaptation strategies and actions to reduce climate change risks. A range of valuable national studies have been carried out and published in Egypt. Under climate change conditions, the productivity of wheat planted at Giza governorate, Egypt in clay soil is expected to decrease by 40 % as an average of 3 cultivars (Khalil et al. 2009). At the same governorate, where wheat was grown in silty clay soil under sprinkler irrigation, the productivity was reduced by 21 % as an average over 4 cultivars (Abdrabbo 2011).

The productivity of wheat planted in El-Behira governorate, Egypt in clay soil under sprinkler irrigation is expected to be reduce by 21 % as an average over four cultivars (Abdrabbo et al. 2013). Under sprinkler irrigation in the same governorate and in sandy soil, wheat productivity was reduced by 30 % for farmer's irrigation (characterized by large applied irrigation water and application of fertilizer and pesticide by broadcasting on the soil). Furthermore, yield losses under applying fertilizer and pesticide via irrigation water through the sprinkler (chemigation) reduced to 25 % and increased water productivity (Ouda et al. 2010). In the same governorate and in sandy soil under sprinkler irrigation, wheat yield was reduced 38 % for farmer's irrigation and by 27 % when irrigation was applied using 1.2 ETc and fertigation application in 80 % of irrigation time (Ibrahim et al. 2012).

In salt-affected soil of Demiatte governorate and under surface irrigation, wheat yield was reduced by 40 % for farmer's irrigation, whereas when irrigation was applied using 1.2 ETc, the reduction in yield was 37 %. However, when wheat was planted on raised bed, the losses in yield became 27 % (Ouda et al. 2012). Wheat yield loss under climate change average over in all these experiments was 32 % (Ouda et al. 2013).

As a consequence for climate change effect, the production–consumption gap for wheat will increase, as well as food insecurity. Therefore, it is important to think about options to increase wheat productivity under current weather conditions, which accordingly will reduce wheat vulnerability under climate change. It has previously been estimated that growing wheat on raised bed could reduce applied water by 20 %, and productivity in tons per fully irrigated hectare can increase by 15 % (Abouelenein et al. 2009). Changing the irrigation system from surface to sprinkler, where application efficiency increases from 60 to 80 %, could also enable farmers to save 20 % of irrigation water for wheat (Taha 2012). As a result, wheat productivity in tons per irrigated hectare will increase by 18 %.

Furthermore, another option can be used to increase wheat cultivated area, i.e., relay intercropping cotton on wheat. This procedure has started to spread among farmers in Egypt (Zohry 2005), where cotton is planted after wheat has reached its reproductive stage, but before it is ready for harvest (Vandermeer 1992). Relay intercropping cotton on wheat saves the first two irrigations applied to cotton (Zohry 2005). Under this system, wheat productivity is reduced by 15 %, and cotton yield is not reduced, compared to sole cultivation (Zohry 2005; El-Bana and Samira 2006; Toaima et al. 2007; Sultan et al. 2012), but wheat cultivated area will increase by the area assigned to cultivate cotton, which compensate for the reduction in productivity.

In this chapter, we investigated options to reduce the applied water for wheat, such as cultivating wheat on raised beds and irrigating wheat with sprinkler system. These two options, if implemented, the national production of wheat, could increase as a result of investing this amount of irrigation water to cultivate new lands. Furthermore, we also investigated the effect of these two options on wheat national production under climate change in 2040. The effect of relay intercropping cotton on wheat on wheat national production was also explored under current climate and under climate change. We also examined the effect of these options on water productivity under current climate and climate change in 2040.

Current Situation of Wheat Production

The cultivated area of wheat, productivity, and total production in 2011/12 growing season were obtained from Ministry of Agriculture and Land Reclamation, Economic Affairs Sector (Table 3.1). Water requirements per hectare were calculated by BISm model (Snyder et al. 2004) and included in Table 3.1. The data revealed that 5,992,160,042 m^3 was used to irrigate 1,109,570 ha and produced 7,398,437 ton of wheat. The total wheat amount that is needed to feed the Egyptians is 12,000,000 ton. Thus, there is a need to increase production through increasing productivity and the cultivated area.

Table 3.1 Wheat cultivated area, productivity, total production, water requirements per hectare, and total water requirements in the studied governorates

Governorates	Cultivated area (ha)[a]	Productivity (ton/ha)[a]	Total production (ton)[a]	Water requirements (m³/ha)[b]	Total water requirements (m³)
Nile Delta					
Alexandria	26,231	6.24	163,680	4617	121,099,014
Demiatte	13,328	6.24	83,164	4950	65,971,125
Kafr El-Sheik	99,833	6.48	646,920	5100	509,150,000
El-Dakahlia	125,010	6.72	840,070	5183	647,970,660
El-Behira	132,316	6.72	889,165	5017	663,786,521
El-Gharbia	61,051	6.72	410,262	4900	299,149,083
El-Monofia	46,192	7.68	354,752	5200	240,196,667
El-Sharkia	168,326	6.00	1,009,958	5600	942,627,000
El-Kalubia	20,674	6.48	133,969	5800	119,910,167
Middle Egypt					
Giza	89.58	7.44	666.50	5733	513,611
Fayoum	68,171	6.72	458,108	5100	347,671,250
Beni Swief	54,425	6.96	378,798	5433	295,709,167
El-Minia	89,075	7.44	662,721	5317	473,584,299
Upper Egypt					
Assuit	69,315	6.96	482,435	6250	433,221,354
Suhag	77,502	6.72	520,811	5467	423,675,778
Qena	39,274	6.72	263,922	6550	257,245,792
Aswan	18,757	5.28	99,035	8033	150,678,556
Total	1,109,570		7,398,437		5,992,160,042

[a]*Source* Central Administration for Agricultural Economics, 2012
[b]Calculated with BISm model

Potential Wheat Productivity Under Raised Beds Cultivation

Cultivating wheat on raised beds could increase total production to 8,508,202 ton (Table 3.2). Furthermore, as a result of reduction in the applied water to raised bed, a large irrigation amount could be obtained, i.e., 1,198,432,008 m³, and could be used to cultivate new area under sprinkler irrigation, i.e., 295,885 ha. The productivity of the new cultivated area is usually 15 % less than its counterpart of the old land. Thus, the total production from old and new land will be 10,185,181 ton, with 38 % increase than its counterpart under surface irrigation.

Table 3.2 Potential wheat production under raised bed, available water for cultivating new land, and total production in the studied governorates

Governorates	Total production (old land) (ton)	Total water requirements (old land) (m³)	Available water for new land (m³)	New area for wheat (ha)	Total production (old + new areas) (ton)
Nile Delta					
Alexandria	188,232	96,879,211	24,219,803	6,995	225,333
Demiatte	95,638	52,776,900	13,194,225	3,554	114,489
Kafr El-Sheik	743,958	407,320,000	101,830,000	26,622	890,593
El-Dakahlia	966,081	518,376,528	129,594,132	33,336	1,156,496
El-Behira	1,022,540	531,029,217	132,757,304	35,284	1,224,084
El-Gharbia	471,801	239,319,267	59,829,817	16,280	564,793
El-Monofia	407,965	192,157,333	48,039,333	12,318	488,375
El-Sharkia	1,161,451	754,101,600	188,525,400	44,887	1,390,375
El-Kalubia	154,064	95,928,133	23,982,033	5,513	184,430
Middle Egypt					
Giza	766,48	410,889	102,722	23,89	917,55
Fayoum	526,824	278,137,000	69,534,250	18,179	630,662
Beni Swief	435,618	236,567,333	59,141,833	14,513	521,479
El-Minia	762,129	378,867,439	94,716,860	23,753	912,346
Upper Egypt					
Assuit	554,801	346,577,083	86,644,271	18,484	664,153
Suhag	598,933	338,940,622	84,735,156	20,667	716,983
Qena	303,511	205,796,633	51,449,158	10,473	363,333
Aswan	113,890	120,542,844	30,135,711	5,002	136,338
Total	8,508,202	4,793,728,033	1,198,432,008	295,885	10,185,181

Potential Wheat Productivity Under Sprinkler Irrigation

Using sprinkler system for irrigation could increase total production to 8,730,155 ton (Table 3.3). As a result of reduction in the applied water for wheat under sprinkler irrigation, a larger irrigation water amount could be obtained, i.e., 1,498,040,010 m³. If this water amount is used to cultivate new area under sprinkler system, this land will be equal to 369,857 ha. Thus, the total production from old and new land will be 10,826,379 ton, which represented 46 % increase than its counterpart under surface irrigation.

Table 3.3 Potential wheat production under sprinkler irrigation, available water for cultivating new land, and total production in the studied governorates

Governorates	Total production (old land) (ton)	Total water requirements (old land) (m^3)	Available water for new land (m^3)	New area for wheat (ha)	Total production (old + new areas) (ton)
Nile Delta					
Alexandria	193,143	90,824,260	30,274,753	8,744	239,519
Demiatte	98,133	49,478,344	16,492,781	4,443	121,696
Kafr El-Sheik	763,366	381,862,500	127,287,500	33,278	946,660
El-Dakahlia	991,283	485,977,995	161,992,665	41,670	1,229,302
El-Behira	1,049,215	497,839,891	165,946,630	44,105	1,301,145
El-Gharbia	484,109	224,361,813	74,787,271	20,350	600,349
El-Monofia	418,607	180,147,500	60,049,167	15,397	519,120
El-Sharkia	1,191,750	706,970,250	235,656,750	56,109	1,477,904
El-Kalubia	158,083	89,932,625	29,977,542	6,891	196,041
Middle Egypt					
Giza	786,470	385,208	128,403	29,86	975,31
Fayoum	540,567	260,753,438	86,917,813	22,724	670,365
Beni Swief	446,982	221,781,875	73,927,292	18,142	554,308
El-Minia	782,011	355,188,224	118,396,075	29,692	969,782
Upper Egypt					
Assuit	569,274	324,916,016	108,305,339	23,105	705,964
Suhag	614,557	317,756,833	105,918,944	25,834	762,120
Qena	311,428	192,934,344	64,311,448	13,091	386,206
Aswan	116,862	113,008,917	37,669,639	6,252	144,922
Total	8,730,155	4,494,120,031	1,498,040,010	369,857	10,826,379

Expected Wheat Production Under Climate Change

A1B climate change scenario was developed using ECHAM5 climate model (Roeckner et al. 2003) in 2040 for each 17 governorates in the Nile Delta and Valley. ECHAM5 model is an Atmospheric Oceanic General Circulation model with low resolution, i.e., 1.9 × 1.9°. Water requirements under climate change were calculated using BISm model (Snyder et al. 2004).

Wheat Grown Under Surface Irrigation

As it was stated previously, water requirements for wheat will be increased under climate change. Ouda et al. (2015) indicated that reduction in the cultivated area of wheat and its total production will occur under climate change. Previous research on the effect of climate change on wheat indicated that its productivity is expected to be reduced by 32 % (Ouda et al. 2012), but technology advances and breeding effort could reduce yield losses percentage to 15 % or could cause no losses. Thus, we calculated wheat production under reduction in the cultivated area and reduction in productivity (32 and 15 %). Moreover, we calculated the production under reduction in the cultivated area only with no yield losses.

Results in Table 3.4 indicated that wheat production will be reduced by 39, 24, and 10 % if wheat yield was reduced by 32, 15, or no yield losses occurred. Thus, if we continue to use surface irrigation to cultivate wheat in 2040, high losses in wheat national production will occur and will increase production–consumption gap and increase food insecurity.

Table 3.4 Expected wheat production under surface irrigation in the studied governorates in 2040

Governorates	Cultivated area (ha)	Production (32 % reduction) (ton)	Production (15 % reduction) (ton)	Production (no yield reduction) (ton)
Nile Delta				
Alexandria	25,467	108,061	135,076	158,913
Demiatte	13,066	55,442	69,303	81,533
Kafr El-Sheik	97,876	431,280	539,100	634,235
El-Dakahlia	122,559	560,047	700,058	823,598
El-Behira	129,722	592,777	740,971	871,731
El-Gharbia	51,738	236,422	295,527	347,679
El-Monofia	38,817	202,715	253,394	298,111
El-Sharkia	142,649	582,009	727,512	855,896
El-Kalubia	17,373	76,553	95,692	112,579
Middle Egypt				
Giza	75,28	380,86	476,071	560,084
Fayoum	59,279	270,881	338,602	398,355
Beni Swief	46,123	218,290	272,863	321,015
El-Minia	74,853	378,698	473,372	556,908
Upper Egypt				
Assuit	59,755	282,807	353,509	415,893
Suhag	65,679	300,128	375,161	441,365
Qena	33,004	150,813	188,516	221,784
Aswan	15,762	56,592	70,739	83,223
Total	993,797	4,503,897	5,629,871	6,623,378

Growing Wheat on Raised Beds Under Climate Change

Changing cultivation method from basin cultivation to raised beds allows reducing in the applied water for wheat and allows using that water amount to irrigate new lands. Water requirements for the new cultivated land will increase under climate change and will have its negative effect on new cultivated area and wheat production. Results in Table 3.5 indicated that under raised beds, wheat production under the assumption that 32 % of yield losses will occur was lower than its counterpart under surface irrigation and current climate by 14 %. Under the assumption of 15 % losses or no yield losses, total production was 9 and 16 % higher than its counterpart under surface irrigation in current climate (Table 3.5). Noreldin et al. (2013) indicated that cultivating wheat on raised beds improves growth environment for wheat and increases its tolerance to stress prevailed by climate change.

Table 3.5 Expected wheat production under raised bed cultivation in the studied governorates in 2040

Governorates	Production (32 % reduction) (ton)	Production (15 % reduction) (ton)	Production (no yield reduction) (ton)
Nile Delta			
Alexandria	176,892	224,745	260,135
Demiatte	76,596	97,166	112,640
Kafr El-Sheik	595,828	755,839	876,217
El-Dakahlia	773,723	981,509	1,137,828
El-Behira	818,941	1,038,870	1,204,325
El-Gharbia	335,829	428,434	493,867
El-Monofia	288,444	368,107	424,182
El-Sharkia	826,725	1,054,692	1,215,772
El-Kalubia	108,928	139,012	160,188
Middle Egypt			
Giza	541,92	691,592	796,944
Fayoum	382,800	487,852	562,942
Beni Swief	310,074	395,576	455,991
El-Minia	538,849	687,671	792,425
Upper Egypt			
Assuit	400,341	510,383	588,737
Suhag	426,323	543,880	626,945
Qena	214,592	273,858	315,576
Aswan	80,524	102,764	118,418
Total	6,355,951	8,091,051	9,346,986

Wheat Irrigated with Sprinkler System

Growing wheat under sprinkler system reduces its vulnerability to climate change and reduced yield losses (Taha 2012). Results in Table 3.6 revealed that wheat production is expected to increase in 2040, if we use sprinkler system for irrigation, compared to its counterpart under surface irrigation. Wheat production under the assumption that 32 % of yield losses will occur in the future was lower than its counterpart under surface irrigation and current climate by 9 %. Under the assumption of 15 % losses or no yield losses, total production was 16 and 34 % higher than its counterpart under surface irrigation in current climate (Table 3.6).

Table 3.6 Expected wheat production under sprinkler system irrigation in the studied governorates in 2040

Governorates	Production (32 % reduction) (ton)	Production (15 % reduction) (ton)	Production (no yield reduction) (ton)
Nile Delta			
Alexandria	159,199	204,653	234,116
Demiatte	81,497	104,879	119,848
Kafr El-Sheik	633,953	815,838	932,284
El-Dakahlia	823,231	1,059,422	1,210,634
El-Behira	871,342	1,121,336	1,281,386
El-Gharbia	360,008	455,576	529,423
El-Monofia	309,351	391,072	454,927
El-Sharkia	886,245	1,121,509	1,303,302
El-Kalubia	116,823	147,684	171,799
Middle Egypt			
Giza	581,201	734,74	854,707
Fayoum	409,798	520,185	602,644
Beni Swief	332,398	420,637	488,821
El-Minia	577,905	730,571	849,861
Upper Egypt			
Assuit	428,773	543,710	630,548
Suhag	457,016	578,336	672,082
Qena	230,145	290,943	338,449
Aswan	86,361	109,175	127,001
Total	6,764,626	8,616,260	9,947,979

Water Requirements for Wheat Under Current and Climate Change

Figure 3.1 presents water requirements per hectare for wheat under surface irrigation, raised beds, and sprinkler system, as an average for all governorates. Water requirements under raised beds and sprinkler system were 20 and 25 % lower than its counterpart under surface irrigation, respectively. This was true under both current climate and climate change conditions. These results implied that there is substianal water amount per hectare which can be saved, if we change water management on field level.

The potentially available water to irrigate new lands after changing water management practice can be reduced under climate change by 50 and 38 % for raised beds and sprinkler irrigation, compared to its values under current climate, respectively (Fig. 3.2). Less availability of water under climate change will reduce the new cultivated area.

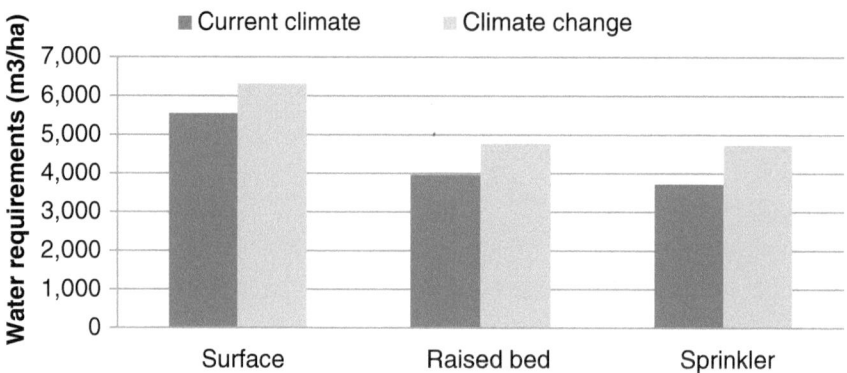

Fig. 3.1 Average water requirements per hectare for wheat under current climate and climate change conditions

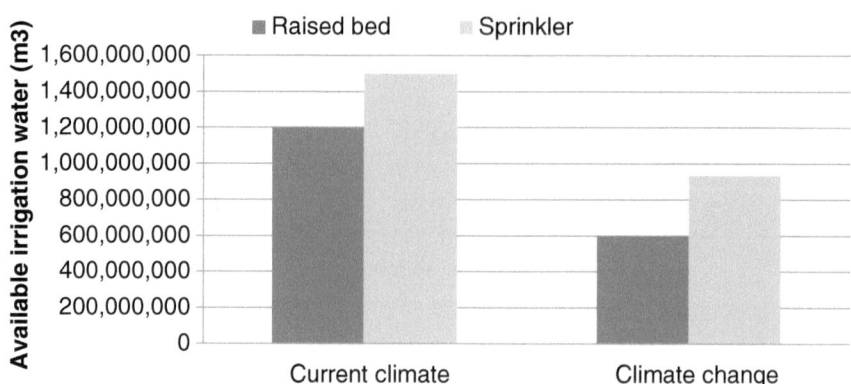

Fig. 3.2 Potentially available water to irrigate new lands under current climate and climate change conditions

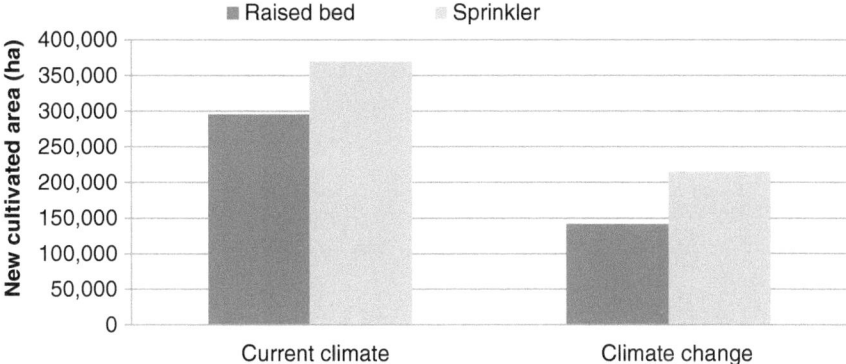

Fig. 3.3 Comparison between potential new cultivated area with wheat under raised beds and sprinkler irrigation

The potential new areas that can be cultivated as a result of reduction of the applied water for wheat under current climate and under climate change in 2040 are presented in Fig. 3.3. The data in the figure implies that under climate change, the potential new cultivated areas will be reduced by 52 and 42 % under raised bed and sprinkler irrigation, respectively, compared to its values under current climate.

Effect of Relay Intercropping Cotton on Wheat

Wheat is cultivated in November and harvested in April. Cotton is cultivated in March. Therefore, if we use this intercropping system, cotton will share its first two irrigations with wheat last two irrigations. Thus, applied water for cotton will be reduced. Under this system, wheat should be cultivated on raised beds in the middle, with 75 % of plant density. Cotton should be cultivated on both sides of the raised bed with 100 % of plant density. As a result, wheat yield will be 15 % less and cotton yield will be the same (Zohry 2005). Thus, the total cultivated area of wheat will consist of wheat area, cotton area, and the added area as a result of water availability from raised bed cultivation. Thus, the cultivated wheat area will increase from 1,109,570 ha (Table 3.1) under current climate to 1,542,068 ha and total wheat production will be 11,565,091 ton (Table 3.7).

Under climate change, the cultivated area will be reduced to 1,166,334 ha and wheat production will be reduced by 32 %. Thus, wheat total production will be reduced to 6,043,115 ton (Table 3.7). Using advanced technology and breeding effort could result in 15 % yield reduction under climate change. Thus, the total wheat production will be 7,466,270 ton. However, if no yield losses occurred under climate change, the production will be 9,459,700 ton (Table 3.7). Noreldin et al. (2013) and Ouda et al. (2012) indicated that cultivation of wheat or cotton on raised

Table 3.7 Total wheat cultivated area and total production under relay intercropping cotton on wheat under current climate and climate change in 2040

Governorates	Current climate		Climate change conditions			
	Cultivated area (ha)	Production (ton)	Cultivated area (ha)	Production (32 % reduction) (ton)	Production (15 % reduction) (ton)	Production (no reduction) (ton)
Nile Delta						
Alexandria	35,133	244,726	34,079	166,414	208,017	220,253
Demiatte	19,784	134,597	19,389	91,526	114,408	121,138
Kafr El-Sheik	175,856	1,199,351	172,339	815,559	1,019,448	1,079,416
El-Dakahlia	177,310	1,312,417	173,763	892,444	1,115,555	1,181,176
El-Behira	218,886	1,567,411	214,508	1,065,839	1,332,299	1,410,670
El-Gharbia	84,005	626,164	68,884	425,792	532,240	563,548
El-Monofia	57,694	503,156	46,732	342,146	427,682	452,840
El-Sharkia	227,874	1,522,373	186,856	1,035,214	1,294,017	1,370,136
El-Kalubia	25,348	187,400	20,532	127,432	159,290	168,660
Middle Egypt						
Giza	108	918	87	624	780	826
Fayoum	91,592	686,566	77,853	466,865	583,581	617,909
Beni Swief	69,702	547,460	57,155	372,273	465,341	492,714
El-Minia	108,220	920,754	87,658	626,113	782,641	828,679
Upper Egypt						
Assuit	86,856	685,911	72,959	466,420	583,024	617,320
Suhag	94,063	723,043	77,132	491,669	614,586	650,739
Qena	47,129	363,333	38,174	247,067	308,833	327,000
Aswan	22,508	136,338	18,231	92,710	115,888	122,705
Total	1,542,068	11,361,919	1,366,334	7,726,105	9,657,631	10,225,727

bed reduces yield vulnerability to climate change and yield losses, as a result of improvement in field growing conditions for either crops.

Thus, relay intercropping cotton on wheat will increase wheat production by 56 %, due to increase in its cultivated area (Fig. 3.4). Furthermore, if losses were 32 %, wheat production will be highly reduced under sole wheat cultivation and under cotton on wheat intercropping, compared to production under current climate. Even if technological advances and breeding efforts were successful to cause no yield losses, wheat potential production under climate change will be lowered by 18 % under relay intercropping, but it still higher than sole wheat production by 28 % under climate change (Fig. 3.5). Furthermore, relay intercropping cotton on wheat will reduce yield losses climate change, compared to sole wheat production under surface irrigation, raised beds, and sprinkler system (Fig. 3.4).

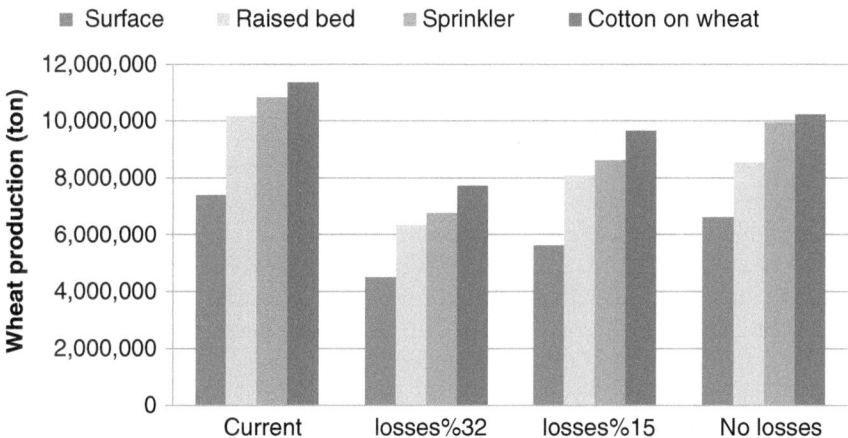

Fig. 3.4 Comparison between potential sole wheat production and potential wheat production using intercropping under current climate and under climate change

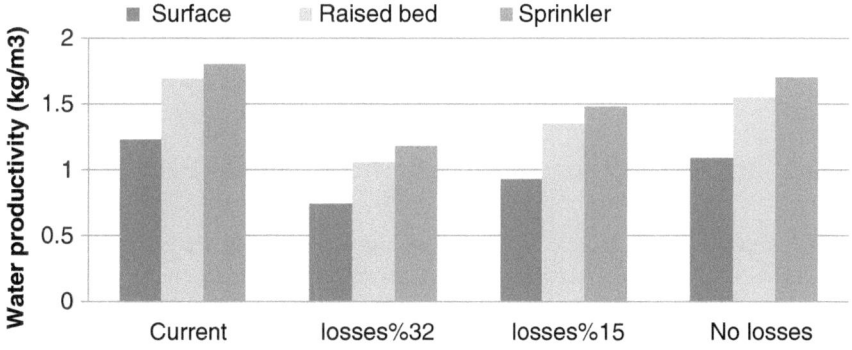

Fig. 3.5 Potential water productivity under current and expected climate change in 2040

Water Productivity Under Current Climate and Under Climate Change

Achieving greater water productivity became the primary challenge for scientists in agriculture. Crop water productivity is a quantitative term used to define the relationship between crop produced and the amount of water involved in crop production. It is a useful indicator for quantifying the impact of irrigation scheduling decisions, with regard to water management (FAO 2003). Optimizing agronomic practices, such as appropriate cropping patterns, high yielding cultivars, irrigation, and fertilization managements, can increase water productivity (Oweis and Hachum 2003). This should include the employment of techniques and practices that deliver more accurate supply of water to crops (Tariq et al. 2003). Hence, the efficiency of water use in agriculture needs to increase in a sustainable manner, i.e., food production (quantitatively and qualitatively) per unit of water used has to be raised (Oweis and Hachum 2003).

Water Productivity for Wheat Under Surface Irrigation

Under current climate and surface irrigation, water productivity was the highest in El-Monofia governorate, i.e., 1.48 kg/m^3 (Table 3.8). This is a result of high productivity for wheat in this governorate (Table 3.1). The lowest water productivity was found in Aswan governorate, i.e., 0.66 kg/m^3, as a result of low productivity and high applied water (Table 3.1). Under climate change, high losses in water productivity will occur, even if no yield losses occurred in 2040. Loss in the average water productivity value was observed under climate change, where it was 40 and 24 % reduction under 32 and 15 % potential reduction in wheat productivity. However, under the assumption that no yield losses will occur, loss in water productivity will be 11 % (Table 3.8). Mahmoud (2014) concluded that water productivity for wheat was reduced under climate change in El-Kalubia governorates from 1.01 kg/m^3 under current climate to 0.59 kg/m^3. Khalil et al. (2009) indicated that water productivities for wheat in Giza governorate were 1.34 and 0.74 kg/m^3 under current climate and climate change, respectively.

Water Productivity for Wheat Grown on Raised Beds

Similar trend for water productivity will be observed if wheat was planted on raised beds (Table 3.9). However, the values of current water productivity were higher under raised bed compared to its counterpart under surface irrigation. Karrou et al. (2012) concluded that water productivity for wheat planted in El-Monofia governorate on raised beds was 1.88 kg/m^3.

Table 3.8 Current and potential wheat water productivity (WP) under surface irrigation in the studied governorates in 2040

Governorates	Current WP (kg/m^3)	Potential WP in 2040 (32 % losses) (kg/m^3)	Potential WP in 2040 (15 % losses) (kg/m^3)	Potential WP in 2040 (no losses) (kg/m^3)
Nile Delta				
Alexandria	1.35	0.89	1.12	1.31
Demiatte	1.26	0.84	1.05	1.24
Kafr El-Sheik	1.27	0.85	1.06	1.25
El-Dakahlia	1.30	0.86	1.08	1.27
El-Behira	1.34	0.89	1.12	1.31
El-Gharbia	1.37	0.79	0.99	1.16
El-Monofia	1.48	0.84	1.05	1.24
El-Sharkia	1.07	0.62	0.77	0.91
El-Kalubia	1.12	0.64	0.80	0.94
Middle Egypt				
Giza	1.30	0.74	0.93	1.09
Fayoum	1.32	0.78	0.97	1.15
Beni Swief	1.28	0.74	0.92	1.09
El-Minia	1.40	0.80	1.00	1.18
Upper Egypt				
Assuit	1.11	0.65	0.82	0.96
Suhag	1.23	0.71	0.89	1.04
Qena	1.03	0.59	0.73	0.86
Aswan	0.66	0.38	0.47	0.55
Average	1.23	0.74	0.93	1.09

Under climate change in 2040, the average water productivity value will be reduced, i.e., 37 and 20 % reduction under 32 and 15 % potential reduction in wheat yield. However, under the assumption that no yield losses will occur under climate change, loss in water productivity will be 8 % (Table 3.9). Noreldin et al. (2013) indicated that water productivity for wheat grown on raised beds was reduced in Demiatte governorate from 1.68 to 1.09 kg/m^3.

Water Productivity for Wheat Irrigated with Sprinkler System

Irrigating wheat with sprinkler system increases water productivity under current climate and reduces the loss in water productivity under climate change (Table 3.10). Loss in the average water productivity value was lower under climate

Table 3.9 Current and potential wheat water productivity (WP) under raised beds in the studied governorates in 2040

Governorates	Current WP (kg/m^3)	Potential WP in 2040 (32 % losses) (kg/m^3)	Potential WP in 2040 (15 % losses) (kg/m^3)	Potential WP in 2040 (no losses) (kg/m^3)
Nile Delta				
Alexandria	1.86	1.46	1.86	2.15
Demiatte	1.74	1.16	1.47	1.71
Kafr El-Sheik	1.75	1.17	1.48	1.72
El-Dakahlia	1.78	1.19	1.51	1.76
El-Behira	1.84	1.23	1.57	1.81
El-Gharbia	1.89	1.12	1.43	1.65
El-Monofia	2.03	1.20	1.53	1.77
El-Sharkia	1.48	0.88	1.12	1.29
El-Kalubia	1.54	0.91	1.16	1.34
Middle Egypt				
Giza	1.79	1.06	1.35	1.55
Fayoum	1.81	1.10	1.40	1.62
Beni Swief	1.76	1.05	1.34	1.54
El-Minia	1.93	1.14	1.45	1.67
Upper Egypt				
Assuit	1.53	0.92	1.18	1.36
Suhag	1.69	1.01	1.28	1.48
Qena	1.41	0.83	1.06	1.23
Aswan	0.90	0.53	0.68	0.79
Average	1.69	1.06	1.35	1.55

change when wheat was irrigated with sprinkler system (Table 3.10). Percentages of reduction in average water productivity were 34, 18, and 6 % under 32, 15 %, and no losses in wheat yield, respectively. Ouda et al. (2010) indicated that water productivity for wheat planted in El-Behira governorate was reduced under climate change by 29 %. Ibrahim et al. (2012) reported that water productivity for wheat was 1.92 kg/m^3 under current climate and 1.22 kg/m^3 under climate change in El-Behira governorate.

Figure 3.5 illustrates that the average water productivity values under surface irrigation will drastically reduced under climate change, compared to its value under current climate. On the contrary, water productivity for wheat under either raised beds cultivation or sprinkler system and 15 % or no yield losses will be higher than its counterpart under surface irrigation in the current climate.

Table 3.10 Current and potential wheat water productivity (WP) under sprinkler system in the studied governorates in 2040

Governorates	Current WP (kg/m^3)	Potential WP in 2040 (32 % losses) (kg/m^3)	Potential WP in 2040 (15 % losses) (kg/m^3)	Potential WP in 2040 (no losses) (kg/m^3)
Nile Delta				
Alexandria	1.9	1.31	1.69	1.93
Demiatte	1.84	1.25	1.59	1.83
Kafr El-Sheik	1.86	1.25	1.61	1.84
El-Dakahlia	1.90	1.27	1.64	1.88
El-Behira	1.96	1.31	1.69	1.94
El-Gharbia	2.01	1.21	1.52	1.78
El-Monofia	2.16	1.29	1.64	1.89
El-Sharkia	1.57	0.96	1.19	1.39
El-Kalubia	1.63	0.98	1.24	1.43
Middle Egypt				
Giza	1.90	1.13	1.43	1.66
Fayoum	1.93	1.18	1.50	1.73
Beni Swief	1.87	1.13	1.43	1.65
El-Minia	2.05	1.22	1.54	1.79
Upper Egypt				
Assuit	1.63	0.99	1.26	1.46
Suhag	1.80	1.08	1.37	1.59
Qena	1.50	0.89	1.13	1.32
Aswan	0.96	0.57	0.72	0.84
Average	1.80	1.18	1.48	1.70

Conclusion

Climate change will increase water requirements for wheat in Egypt, which consequently will reduce its cultivated area and production. Since wheat is an important cereal crop in Egypt, where its consumption is increasingly rise from one year to another due to the population increase, such shortage is compensated through the importation, adding a heavy burden on the country's national budget. The appropriate way to overcome such existing gap is to increase the production through increasing the irrigated area. However, with the current prevailing irrigation system, i.e., surface irrigation, it will be impossible to achieve such approach. Thus, reduction of the applied water for wheat through cultivation on raised beds or replacing surface irrigation with sprinkler system could lead to save on the applied water, increase new cultivated area with wheat, increase national production, and increase crop water productivity. Under climate change, surface irrigation will cause a lot of losses in wheat production, where productivity will be low, water requirements will be high, and water productivity will be low. The production–consumption gap will increase and food insecurity will increase.

Cultivation on raised beds does not include any extra cost, in fact, it reduces variable costs, as it requires less fertilizers and less fuel to operate irrigation pump, whereas implementation of sprinkler system for wheat irrigation includes extra cost for farmers and they might not be able to pay it. Therefore, the government of Egypt should bear this cost on behalf of the farmers to face water scarcity and reduce food gap.

References

Abdrabbo M, Ouda S, Noreldin T (2013) Modeling the effect of irrigation scheduling on wheat under climate change conditions. Nat Sci J 115:10–18

Abdrabbo M (2011) Water management for some field crops grown under climate change conditions: a project report. Sci Technol Dev Fund, Egypt

Abouelenein R, Oweis T, El Sherif M, Awad H, Foaad F, Abd El Hafez S, Hammam A, Karajeh F, Karo M, Linda A (2009) Improving wheat water productivity under different methods of irrigation management and nitrogen fertilizer rates. Egypt J Appl Sci 24(12A):417–431

Bukhat NM (2005) Studies in yield and yield associated traits of wheat Triticum aestivum L. genotypes under drought conditions. MSc thesis, Department of Agronomy, Sindh Agriculture University, Tandojam, Pakistan

Dencic S, Kastori R, Kobiljski B, Duggan B (2000) Evaporation of' grain yield and its components in wheat cultivars and land races under near optimal and drought conditions. Euphytica. 1:43–52

El-Bana H, Samira MA (2006) Performance of cotton and wheat under intercropping, planting patterns and nitrogen fertilization. Ann Agric Sci Moshtohor 444:1385–1404

FAO (2003) Optimizing soil moisture for plant production—the significance for soil porosity. F. Shaxson & R. Barber. FAO Soils Bulletin 79. Rome

Gupta NK, Gupta S, Kumar A (2001) Effect of water stress on physiological attributes and their relationship with growth and yield in wheat cultivars at different growth stages. J Agron 86 (143):7–1439

Ibrahim M, Ouda S, Taha A, El Afandi G, Eid SM (2012) Water management for wheat grown in sandy soil under climate change conditions. J Soil Sci Plant Nutr 122:195–210

Karrou M, Oweis T, El Enein R, Sherif M (2012) Yield and water productivity of maize and wheat under deficit and raised bed irrigation practices in Egypt. Afr J Agric Res 711:1755–1760

Khalil FA, Farag H, El Afandi G, Ouda SA (2009) Vulnerability and adaptation of wheat to climate change in Middle Egypt. In: 13th conference on water technology, Hurghada, Egypt, 12–15 Mar 2009

Mahmoud I (2014) Simulation of the effect of adaptation strategies on improving yield of some crops grown under climate change conditions. MSc, Institute of Environmental Research, Ain Sham University, Cairo

McMaster GS (1997) Phonology, development, and growth of wheat (Triticumaestivum L.) shoot apex: a review. Adv Agron 59:63–118

Noreldin T, Ouda S, Abou Elenein R (2013) Development of management practices to address wheat vulnerably to climate change in North Delta. In: Proceeding of the 11th international conference on development of dry lands, pp 982–995

Ouda SA, Noreldin T, Abd El-Latif K (2015) Water requirements for wheat and maize under climate change in North Nile Delta. Span J Agric Res 13(1):10

Ouda S, Khalil FA, Noreldin T (2013) Modeling climate change impacts and adaptation strategies for crop production in Egypt: an overview. Springer Publishing House, New York

Ouda S, Noreldin T, Abou Elenin R, Abd El-Baky H (2012) Improved agricultural management practices reduced wheat vulnerablyility to climate change in salt affected soils. Egypt J Agric Res 904:499–513

Ouda S, Sayed M, El Afandi G, Khalil F (2010) Developing an adaptation strategy to reduce climate change risks on wheat grown in sandy soil in Egypt. In: 10th international conference on development of dry lands. Cairo, Egypt, 12–15 Dec 2010

Oweis T, Hachum AY (2003) Improving water productivity in the dry areas of West Asia and North Africa. In: Kijne JW, Barker R, Molden D (ed) Water productivity in agriculture: limits and opportunities for improvement

Roeckner E, Bäuml G, Bonaventura L, Brokopf R, Esch M, Giorgetta M, Hagemann S, Kirchner I, Kornblueh L, Manzini E, Rhodin A, Schlese U, Schulzweida U, Tompkins A (2003) The atmospheric general circulation model ECHAM5. Part I: Model Description. MPI Report 349, Max Planck Institute for Meteorology, Hamburg, Germany, pp 127

Slafer GA, Satorre EH (1999) Wheat: ecology and physiology of yield determination. Haworth Press Technology and Industrial, New York. ISBN 1560228741

Snyder RL, Orang M, Bali K, Eching S (2004) Basic irrigation scheduling BISm. http://www.waterplan.water.ca.gov/landwateruse/wateruse/Ag/CUP/Californi/Climate_Data_010804.xls

Sultan MS, El-Kassaby AT, Ghonema MH, Ogeaz AA, Abd-Allah AM (2012) Rely intercropping wheat and cotton studies: II-Effect of sowing sates and ridge width on cotton. J Biol Sci 12(6):349–354. doi:10.3923/jbs. ISSN:1727-3048

Taha A (2012) Effect of climate change on maize and wheat grown under fertigation treatments in newly reclaimed soil. Ph.D. thesis, Tanta University, Egypt

Tariq JA, Khan MJ, Usman K (2003) Irrigation Scheduling of maize crop by pan evaporation method. Pakistan J Water Res 72:29–33

Toaima SC, Gadallah RA, Mohamadin ESA (2007) Effect of relay intercropping systems on yield and its components of wheat and cotton. Ann Agric Sci Ain Shams Univ Cairo 522:317–326

Vandermeer JH (1992) The ecology of intercropping. Cambridge University Press, Cambridge, p 237

Wollenweber B, Porter JR, Schellberg J (2003) Lack of interaction between extreme high-temperature events at vegetative and reproductive growth stages in wheat. J Agron Crop Sci 189:142–150

Zohry AA (2005) Effect of relaying cotton on some crops under bio-mineral N fertilization rates on yield and yield components. Ann Agric Sci 431:89–103

Chapter 4
Combating Adverse Consequences of Climate Change on Maize Crop

Tahany Noreldin, Samiha Ouda and Ahmed Taha

Abstract In this chapter we investigated the effect of cultivating maize on raised beds and irrigation with drip system on increasing national maize production through increasing productivity, reducing the applied irrigation water and use it to irrigate more land with maize. Under climate change, maize vulnerability can be reduced by the above practices. We also calculated the contribution of each option in reducing maize production-consumption gap under current climate and under climate change in 2040. The effect of these practices on water and land productivity under preset time and under climate change in 2040 was also examined. The results revealed that production-consumption gap in maize are about 45 %. The results also indicate that cultivating maize on raised beds or using drip system for irrigation reduced production-consumption gap under current climate and in 2040, where the percentage of imported maize will reduce to 23 and 12 % under both systems, respectively, under current climate and will reduce yield losses under climate change. The results also indicate that water productivity was the lowest under surface irrigation and was the highest when drip system was used under both current and climate change.

Keywords Surface irrigation · Raised beds cultivation · Drip system · Maize production-consumption gap · Water productivity

Most recent assessments of the effect of climate change on arid and semi-arid regions concluded that these areas are highly vulnerable to climate change. The projected climatic changes will be among the most important challenges for agriculture in the twenty-first century, especially for developing countries and arid regions (IPCC 2007). The risks associated with agriculture and climate change arise out strong complicated relationships between agriculture and the climate system, plus the high reliance of agriculture on finite natural resources (Abou-Zeid 2002). The inter annual, monthly and daily distribution of weather variables, such as temperature, radiation, precipitation, water vapor pressure and wind speed affects a

T. Noreldin (✉) · S. Ouda · A. Taha
Water Requirement and Field Irrigation Department, Soils, Water and Environment Research Institute, Agricultural Research Center, Giza, Egypt

© The Author(s) 2016
S. Ouda, *Major Crops and Water Scarcity in Egypt*,
SpringerBriefs in Water Science and Technology,
DOI 10.1007/978-3-319-21771-0_4

number of physical, chemical and biological processes that drive the productivity of agricultural (IPCC 2007).

Maize is planted in Egypt as a summer crop. It is important to the national economy, both as a source of human food and feed. Maize production in Egypt has significantly increased over the past three decades. The cultivated area of maize in 2012 was 679,508 hectare with average productivity equal 6.87 ton/ha. There is a gap between production and consumption of maize in Egypt estimated by around 45 %. The abiotic stress that climate change will cause, i.e. water and heat stress can disturb physical and chemical processes in maize. Drought and temperature extremes can cause extensive economic loss to agriculture (Peng et al. 2004). The total yield loss depends on when the stress occurs (growth stage), as well as the duration and the severity of the stress. Early season drought reduces plant growth and inhibits plant development (Heiniger 2001). Drought occurring between 2 weeks before and 2 weeks after silking stage can cause significant reductions in kernel set and kernel weight (Schussler and Westgate 1991), resulting in an average of 20–50 % yield loss (Nielson 2007). High temperature stress at critical developmental stages of maize plants also causes significant yield loss (Lobell et al. 2011). Maize plants become susceptible to high temperatures after reaching eight-leaf stage to seventeen-leaf stages (Chen et al. 2010), which has significant impact on plant growth, architecture, ear size and kernel numbers (Farré and Faci 2006).

Previous research on the effect of climate change on maize revealed that reduction in maize yield planted in clay soil under surface irrigation by 55 % is expect to occur under climate change in Giza governorate, with deterioration in water productivity (Ouda et al. 2009). In the same governorate, maize yield grown in silty clay soil under drip irrigation, where its yield was reduced by 27 % as an average of 4 hybrids (Ouda et al. 2012a, b).

The productivity of maize planted in El-Behira governorate in clay soil under drip irrigation is expected to be reduced by 25 % (Ouda et al. 2012a, b). In the same governorate and in sandy soil under drip irrigation, maize yield was reduced 41 % for farmer's irrigation and by 36 % when irrigation was applied using 1.0 ETc and fertigation application in 80 % of irrigation time (Ouda et al. 2014). Thus, the yield of maize will be highly reduced under future climate change. Maize yield losses under climate change average over all these experiments were 42 % (Ouda et al. 2013). The level of yield reduction depend on geographic location, soil type and irrigation system.

Understanding the potential impacts of climate change is very important in developing adaptation strategies and actions to reduce future climate change risks on maize to reduce production-consumption gap. One of these adaptation strategies is cultivation on raised beds, which could reduce applied water by 20 % and productivity in tons per fully irrigated hectare can increase by 15 % (Abouelenein et al. 2009). Hobbs et al. (2000) demonstrated that raised beds planting contributed significantly to improved water distribution and efficiency, increased fertilizer use efficiency and reduced weed infestation, lodging and seed rate without sacrificing yield. The other option that can be used to reduce applied water to maize is irrigation with drip system, where application efficiency increases from 60 to 90 % and maize productivity will increase by 18 % (Taha 2012).

In this chapter we investigated the effect of cultivation of maize on raised beds and irrigation with drip system on increasing national maize production through increasing productivity, reducing the applied irrigation water and use it to irrigate more land with maize. Under climate change, maize vulnerability can be reduced by the above practices. We also calculated contribution of each option in reducing maize production-consumption gap under current climate and under climate change in 2040. The effect of these practices on water and land productivity under preset time and under climate change in 2040 was also examined.

Current Situation of Maize Production

Table 4.1 presents maize cultivated area, productivity and total production in 2013 growing season. These data were obtained from Ministry of Agriculture and Land Reclamation, Economic Affairs Sector. BISm model (Snyder et al. 2004) was used to calculate water requirements per hectare. The table revealed that $4,618,943,455$ m^3 was used to irrigate 679,508 hectare and produced 5,478,125 ton maize. Since production-consumption gap in maize is about 45 %, there is a need to increase production through increase productivity per hectare and increase cultivated area.

Potential Maize Productivity Under Improved Management Practices

Cultivation on Raised Beds

Table 4.2 indicated that cultivating maize on raised beds could increase total production to 6,299,844 ton. As a result of cultivation on raised beds, an amount of irrigation water estimated by $923,788,691$ m^3 can be obtained and use to cultivated new area equal to 181,202 hectare under drip irrigation. The productivity of the new cultivated area is usually 15 % less than its counterpart of the old land. Thus, the total production from old and new land will be 7,541,552 ton, with 41 % increase than its counterpart under surface irrigation.

Irrigation with Drip System

Irrigating maize with drip system could increase total maize production to 6,464,187 ton (Table 4.3). Furthermore, as a result of reduction in the applied water for maize under drip system, a larger irrigation water amount could be obtained, i.e. $1,539,647,818$ m^3 and can be used to cultivated new area under drip irrigation, i.e.

Table 4.1 Maize cultivated area, productivity, total production, water requirements per hectare and total water requirements in the studied governorates

Governorates	Cultivated area (ha)[a]	Productivity (ton/ha)[a]	Total production (ton)[a]	Water requirements (m³/ha)[b]	Total water requirements (m³)
Nile Delta					
Alexandria	13,243	7.7	102,249	6,431	85,161,991
Demiatte	735	8.8	6,472	6,125	4,504,427
Kafr El-Sheik	23,008	9.0	206,412	6,500	149,554,167
El-Dakahlia	22,725	9.5	215,647	5,861	133,191,308
El-Behira	63,480	9.0	569,336	6,069	385,285,804
El-Gharbia	18,753	8.9	167,111	5,917	110,954,757
El-Monofia	81,058	9.8	792,751	6,111	495,356,482
El-Sharkia	100,376	8.1	812,081	6,361	638,501,829
El-Kalubia	24,799	7.8	193,668	6,681	165,669,427
Middle Egypt					
Giza	20,582	8.6	177,924	7,264	149,502,940
Fayoum	25,898	6.4	165,705	7,333	189,918,056
Beni Swief	56,025	6.9	386,301	7,653	428,743,686
El-Minia	104,357	7.7	798,454	7,514	784,124,398
Upper Egypt					
Assuit	51,005	7.2	365,033	7,125	363,410,625
Suhag	53,386	8.0	427,047	7,069	377,411,129
Qena	16,038	4.3	68,208	7,542	120,955,764
Aswan	4,040	5.9	23,726	9,083	36,696,667
Total	679,508		5,478,125	6,861	4,618,943,455

[a]*Source* Central Administration for Agricultural Economics, 2013
[b]Calculated with BISm model

339,754 hectare. Thus, the total production from old and new land will be 8,792,391 ton, which represented 61 % increase than its counterpart under surface irrigation.

Contribution in Reduction of Production-Consumption Gap

National maize consumption is estimated by approximately 10,000,000 ton in 2013. Using surface irrigation for maize allows production of 5,478,125 ton, i.e. 55 % only. Thus, there is a need to import 4,521,875 ton, which put heavy burden on Egyptian's government to fill that gap. Cultivation on raised beds or using drip system for irrigation can reduce this gap. Figure 4.1 showed that importation of maize could be reduced from 4,521,875 to only 1,207,609 ton if drip irrigation

Table 4.2 Potential maize production under raised beds, available water for cultivating new land and total production in the studied governorates

Governorates	Total production (old land) (ton)	Total water requirements (old land) (m³)	Available water for new land (m³)	New area for maize (ha)	Total production (old + new areas) (ton)
Nile Delta					
Alexandria	117,586	68,129,593	17,032,398	3,532	140,763
Demiatte	7,443	3,603,542	900,885	196	8,910
Kafr El-Sheik	237,374	119,643,333	29,910,833	6,136	284,161
El-Dakahlia	247,994	106,553,046	26,638,262	6,060	296,874
El-Behira	654,736	308,228,644	77,057,161	16,928	783,785
El-Gharbia	192,178	88,763,806	22,190,951	5,001	230,056
El-Monofia	911,663	396,285,185	99,071,296	21,616	1,091,353
El-Sharkia	933,893	510,801,463	127,700,366	26,767	1,117,964
El-Kalubia	222,719	132,535,542	33,133,885	6,613	266,617
Middle Egypt					
Giza	204,613	119,602,352	29,900,588	5,488	244,943
Fayoum	190,561	151,934,444	37,983,611	6,906	228,121
Beni Swief	444,246	342,994,949	85,748,737	14,940	531,807
El-Minia	918,222	627,299,519	156,824,880	27,828	1,099,205
Upper Egypt					
Assuit	419,787	290,728,500	72,682,125	13,601	502,528
Suhag	491,104	301,928,903	75,482,226	14,236	587,902
Qena	78,439	96,764,611	24,191,153	4,277	93,899
Aswan	27,285	29,357,333	7,339,333	1,077	32,663
Total	6,299,844	3,695,154,764	923,788,691	181,202	7,541,552

implemented. Thus, percentage of imported maize will be reduce from 45 % under surface irrigation to 23 and 12 % when cultivation on raised beds or using drip system for irrigation implemented, respectively.

Expected Maize Production Under Climate Change

The ECHAM5 climate model (Roeckner et al. 2003) was used to develop A1B climate change scenario in 2040 for each 17 governorates in the Nile Delta and Valley. The model is an Atmospheric Oceanic General Circulation model with 1.9 × 1.9 ° resolution. Water requirements under climate change were calculated using BISm model (Snyder et al. 2004).

Table 4.3 Potential maize production under drip irrigation, available water for cultivating new land and total production in the studied governorates

Governorates	Total production (old land) (ton)	Total water requirements (old land) (m³)	Available water for new land (m³)	New area for maize (ha)	Total production (old + new areas) (ton)
Nile Delta					
Alexandria	120,654	56,774,660	28,387,330	6,622	164,110
Demiatte	7,637	3,002,951	1,501,476	368	10,388
Kafr El-Sheik	243,567	99,702,778	49,851,389	11,504	331,292
El-Dakahlia	254,464	88,794,205	44,397,103	11,362	346,114
El-Behira	671,816	256,857,203	128,428,601	31,740	913,784
El-Gharbia	197,191	73,969,838	36,984,919	9,376	268,213
El-Monofia	935,446	330,237,654	165,118,827	40,529	1,272,365
El-Sharkia	958,255	425,667,886	212,833,943	50,188	1,303,389
El-Kalubia	228,529	110,446,285	55,223,142	12,399	310,838
Middle Egypt					
Giza	209,951	99,668,627	49,834,313	10,291	285,569
Fayoum	195,532	126,612,037	63,306,019	12,949	265,957
Beni Swief	455,835	285,829,124	142,914,562	28,012	620,013
El-Minia	942,175	522,749,599	261,374,799	52,178	1,281,518
Upper Egypt					
Assuit	430,738	242,273,750	121,136,875	25,503	585,877
Suhag	503,916	251,607,419	125,803,709	26,693	685,411
Qena	80,485	80,637,176	40,318,588	8,019	109,474
Aswan	27,997	24,464,444	12,232,222	2,020	38,080
Total	6,464,187	3,079,295,637	1,539,647,818	339,754	8,792,391

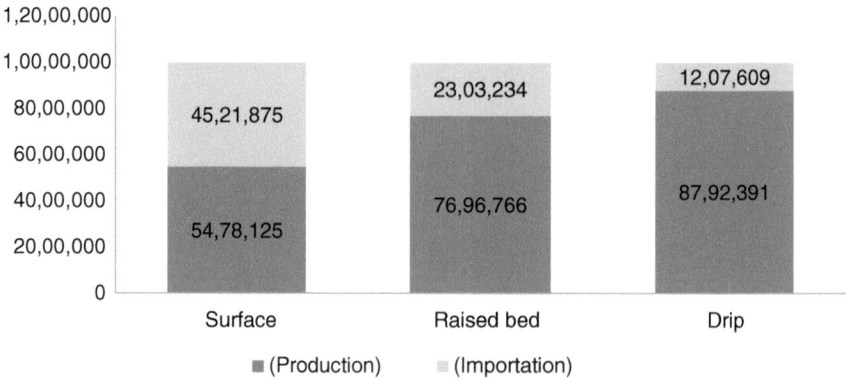

Fig. 4.1 Current and potential maize production and importation under improved management practices

Maize Grown Under Surface Irrigation

Ouda et al. (2015) indicated that water requirements for maize will be increase under climate change, which will result in reduction in the cultivated area and total production. Furthermore, under climate change pervious research indicated that maize yield is expected to reduce by 42 % (Ouda et al. 2013). One of the proposed adaptations to climate change is breading for varieties with high tolerance to water and salinity stresses and with high water use efficiency. Thus, we assumed that technological advances and breeding effort could reduce yield losses percentage to only 15 %. A more pessimistic assumption is that these efforts will be successful to cause no yield losses. Thus, we calculated the maize production under reduction in the cultivated area and reduction in productivity (42 and 15 %). Moreover, we calculated maize production under reduction in the cultivated area only, as for the third assumption, i.e. no yield losses will occur.

Under climate change in 2040, high yield losses are expected to happen under the three assumptions. The potential maize production will be 2,763,357; 4,049,748 and 4,764,409 ton, if maize yield was reduced by 42 and 15 % or no yield losses occur (Table 4.4). These results are very disturbing and indicate that growing maize under surface irrigation in 2040 will cause high losses in maize national production that will increase production-consumption gap, which is already high.

Maize Production-Consumption Gap in 2040

If we assumed that maize consumption in 2040 will increase by 10 % as a result of population growth, the total consumption will be 11,000,000 ton. Thus, the contribution of each of the above assumption in production-consumption gap was calculated and presented in Fig. 4.2. The figure illustrated that the gap between maize production and consumption will increase to 75, 63 and 57 % under the assumption that 42 and 15 % or no yield will occur in 2040.

Growing Maize on Raised Beds Under Climate Change

Changing cultivation method from basin cultivation to raised beds increased maize productivity by 15 % and increase production (Table 4.5). Furthermore, it allows reduction of the applied water for maize, and consequently using that water amount to irrigate new lands under drip irrigation. Maize water requirements for new land will increase under climate change and the new cultivated area will decrease. Thus, the potential maize production will be 3,913,367; 5,847,191 and 6,747,184 ton under the assumption that 42 and 15 % or no yield losses, respectively (Table 4.5). Karrou et al. (2012) indicated that cultivating maize on raised bed improving growth environment for maize and increase its tolerance to prevailed stress.

Table 4.4 Expected maize production under surface irrigation in the studied governorates in 2040

Governorates	Cultivated area (ha)	Production (42 % reduction) (ton)	Production (15 % reduction) (ton)	Production (no yield reduction) (ton)
Nile Delta				
Alexandria	11,902	53,298	78,109	91,893
Demiatte	666	3,398	4,980	5,859
Kafr El-Sheik	20,747	107,955	158,210	186,129
El-Dakahlia	20,044	110,324	161,681	190,213
El-Behira	55,147	286,869	420,412	494,602
El-Gharbia	16,458	85,063	124,662	146,661
El-Monofia	70,327	398,924	584,630	687,800
El-Sharkia	88,248	414,096	606,864	713,958
El-Kalubia	21,501	97,391	142,728	167,916
Middle Egypt				
Giza	17,697	88,730	130,036	152,983
Fayoum	22,362	82,986	121,618	143,079
Beni Swief	48,342	193,332	283,331	333,331
El-Minia	89,907	398,981	584,714	687,899
Upper Egypt				
Assuit	43,936	182,376	267,275	314,441
Suhag	46,132	214,033	313,669	369,022
Qena	13,791	34,018	49,854	58,652
Aswan	3,400	11,583	16,975	19,970
Total	590,609	2,763,357	4,049,748	4,764,409

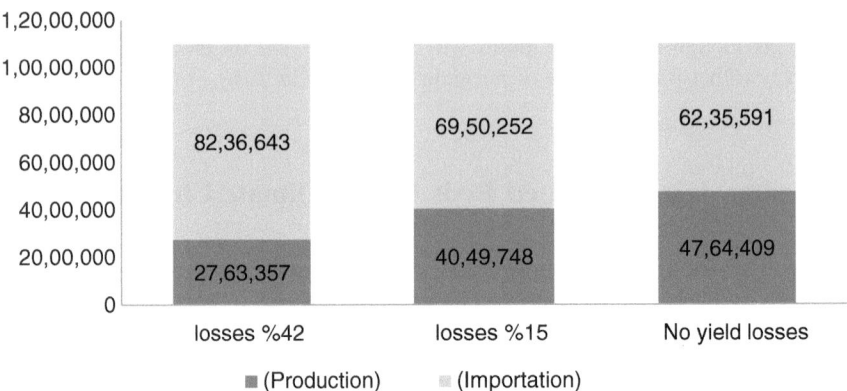

Fig. 4.2 Current and potential maize production and importation in 2040 grown under surface irrigation

Table 4.5 Potential maize production under raised beds cultivation in 2040 in the studied governorates

Governorates	Production (42 % reduction) (ton)	Production (15 % reduction) (ton)	Production (no yield reduction) (ton)
Nile Delta			
Alexandria	83,251	124,288	143,536
Demiatte	4,764	7,108	8,215
Kafr El-Sheik	151,481	226,012	261,173
El-Dakahlia	155,469	232,163	268,049
El-Behira	405,471	605,860	699,087
El-Gharbia	119,990	179,218	206,880
El-Monofia	563,997	842,777	972,409
El-Sharkia	583,920	872,088	1,006,759
El-Kalubia	137,710	205,784	237,430
Middle Egypt			
Giza	125,672	187,859	216,676
Fayoum	117,437	175,519	202,478
Beni Swief	273,629	408,972	471,775
El-Minia	564,867	844,314	973,909
Upper Egypt			
Assuit	258,211	385,953	445,191
Suhag	302,841	452,606	522,140
Qena	48,180	72,021	83,070
Aswan	16,476	24,650	28,406
Total	3,913,367	5,847,191	6,747,184

Contribution of Raised Beds Cultivation in Reduction of Production-Consumption Gap

Cultivation of maize on raised beds can reduce amount of importation by 64, 47 and 39 % under 42 and 15 % or no yield losses, respectively compared to surface irrigation cultivation (Fig. 4.3).

Maize Irrigated with Drip System

Taha (2012) indicated that growing maize under drip system reduces its vulnerability to climate change and reduced yield losses. Results in Table 4.6 revealed that maize national production was increased to 5,007,519; 6,467,611 and 8,542,941 ton, under 42, 15 % or no yield losses compared to its counterpart under surface irrigation.

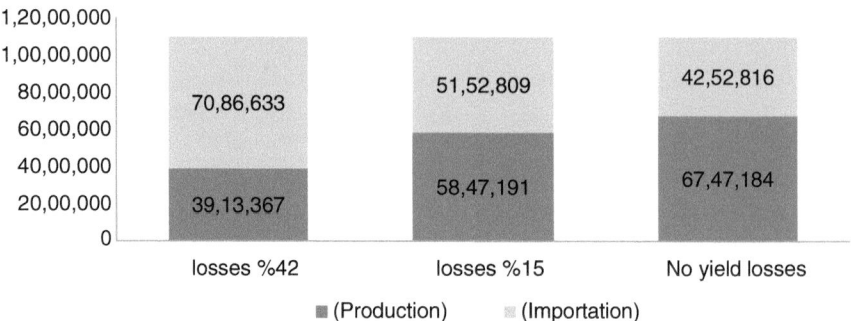

Fig. 4.3 Current and potential maize production and importation under raised beds 2040

Table 4.6 Potential maize production under drip system in the studied governorates in 2040

Governorates	Production (42 % reduction) (ton)	Production (15 % reduction) (ton)	Production (no yield reduction) (ton)
Nile Delta			
Alexandria	121,589	129,638	168,545
Demiatte	7,039	6,514	9,873
Kafr El-Sheik	229,139	216,560	320,486
El-Dakahlia	182,658	195,526	304,257
El-Behira	502,317	572,126	831,785
El-Gharbia	148,393	170,254	245,498
El-Monofia	683,063	812,760	1,134,945
El-Sharkia	688,526	771,981	1,146,727
El-Kalubia	164,202	205,982	273,475
Middle Egypt			
Giza	161,461	212,021	266,306
Fayoum	166,953	270,991	271,558
Beni Swief	382,551	595,581	623,617
El-Minia	791,875	1,187,860	1,290,632
Upper Egypt			
Assuit	305,497	448,336	509,781
Suhag	395,427	508,169	650,385
Qena	57,446	137,104	95,770
Aswan	19,384	26,208	32,464
Total	5,007,519	6,467,611	8,176,102

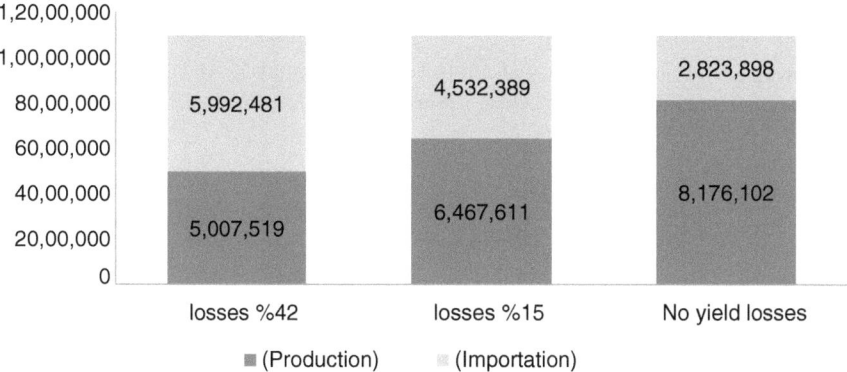

Fig. 4.4 Current and potential maize production and importation in 2040 grown under drip system

Contribution of Irrigation with Drip System in Reducing Maize Production-Consumption Gap

Using drip system to irrigate maize in 2040 could reduce production-consumption gap, compared to its counterpart under surface irrigation. This will reduce importation to 5,992,481 ton, which represent 54 % under the assumption that 42 % yield losses will occur under climate change. Under the assumption that 15 % losses in productivity will occur, the importation will be 4,532,389 ton or 41 % of the total consumption. If no yield losses occur under climate change, the importation will represent 26 % of the total consumption (Fig. 4.4).

Maize Water Productivity

Water Productivity for Maize Under Surface Irrigation

Table 4.7 indicated that under current climate, water productivity was the highest in El-Dakahlia governorate and the lowest was found in Qena, i.e. 1.62 and 0.56 kg/m^3, respectively. The national value of water productivity was reduced from 1.18 to 0.60 and 0.68 kg/m^3 under climate change and 32 and 15 % potential reduction in maize yield, respectively. However, under the assumption that no yield losses will occur, water productivity was 0.80 kg/m^3 (Table 4.7). Mahmoud (2014) concluded that water productivity for maize in El-Kalubia governorate was 1.20 kg/m^3 and was reduced to 0.62 kg/m^3 under climate change. Ouda et al. (2009) indicated that water productivity was 0.88 kg/m^3 in El-Giza governorate for maize grown under surface irrigation. Under climate change, water productivity was reduced to 0.46 kg/m^3.

Table 4.7 Current and potential maize water productivity (WP) under surface irrigation in the studied governorates in 2040

Governorates	Current WP (kg/m^3)	Potential WP in 2040 (42 % losses) (kg/m^3)	Potential WP in 2040 (15 % losses) (kg/m^3)	Potential WP in 2040 (no losses) (kg/m^3)
Nile Delta				
Alexandria	1.20	0.63	0.92	1.08
Demiatte	1.44	0.75	1.11	1.30
Kafr El-Sheik	1.38	0.72	1.06	1.24
El-Dakahlia	1.62	0.83	1.21	1.43
El-Behira	1.48	0.74	1.09	1.28
El-Gharbia	1.51	0.77	1.12	1.32
El-Monofia	1.60	0.81	1.18	1.39
El-Sharkia	1.27	0.65	0.95	1.12
El-Kalubia	1.17	0.59	0.86	1.01
Middle Egypt				
Giza	1.19	0.59	0.87	1.02
Fayoum	0.87	0.44	0.64	0.75
Beni Swief	0.90	0.45	0.66	0.78
El-Minia	1.02	0.51	0.75	0.88
Upper Egypt				
Assuit	1.00	0.50	0.74	0.87
Suhag	1.13	0.57	0.83	0.98
Qena	0.56	0.28	0.41	0.48
Aswan	0.65	0.32	0.46	0.54
Average	1.18	0.60	0.68	0.80

Water Productivity for Maize Grown on Raised Beds

Similar trend for water productivity was observed if maize was planted on raised beds (Table 4.8). However, the values of current water productivity was higher under raised beds compared its counterpart under surface irrigation. Under climate change in 2040, the national water productivity value will be reduce from 1.62 kg/m^3 under current climate to 0.85 and 1.27 kg/m^3 under 42 and 15 % potential reduction in maize yield, respectively. However, under the assumption that no yield losses will occur, water productivity was 1.46 kg/m^3 (Table 4.8). Karrou et al. (2012) indicated that water productivity for maize planted in El-Monofia governorate on raised beds was 1.99 kg/m^3.

Table 4.8 Current and potential maize water productivity (WP) under raised beds in the studied governorates in 2040

Governorates	Current WP (kg/m³)	Potential WP in 2040 (42 % losses) (kg/m³)	Potential WP in 2040 (15 % losses) (kg/m³)	Potential WP in 2040 (no losses) (kg/m³)
Nile Delta				
Alexandria	1.65	0.98	1.46	1.69
Demiatte	1.98	1.06	1.58	1.82
Kafr El-Sheik	1.90	1.01	1.51	1.75
El-Dakahlia	2.23	1.17	1.74	2.01
El-Behira	2.03	1.05	1.57	1.81
El-Gharbia	2.07	1.08	1.62	1.86
El-Monofia	2.20	1.14	1.70	1.96
El-Sharkia	1.75	0.91	1.37	1.58
El-Kalubia	1.61	0.83	1.24	1.43
Middle Egypt				
Giza	1.64	0.84	1.26	1.45
Fayoum	1.20	0.62	0.92	1.07
Beni Swief	1.24	0.64	0.95	1.10
El-Minia	1.40	0.72	1.08	1.24
Upper Egypt				
Assuit	1.38	0.71	1.06	1.23
Suhag	1.56	0.80	1.20	1.38
Qena	0.78	0.40	0.60	0.69
Aswan	0.89	0.45	0.67	0.77
Average	1.62	0.85	1.27	1.46

Water Productivity for Maize Irrigated with Drip System

Irrigating maize with drip system increased water productivity under current climate and reduce the loss in water productivity under climate change (Table 4.9). Loss in the national water productivity value was lower under climate change when maize was irrigated with drip system (Table 4.9). Water productivity was 1.10, 1.36 and 1.75 kg/m³ under 42, 15 % and no losses in maize yield under climate change, respectively. Ouda et al. (2014) indicated that water productivity for maize planted in El-Behira governorate irrigated with drip system was 1.36 kg/m³ under current climate and was reduced to 1.10 under climate change. Similarly, water productivity at El-Giza governorate was 1.44 kg/m³ under current climate and reduced to 1.00 kg/m³ under climate change (Ouda et al. 2009).

Table 4.9 Current and potential maize water productivity (WP) under drip system in the studied governorates in 2040

Governorates	Current WP (kg/m^3)	Potential WP in 2040 (42 % losses) (kg/m^3)	Potential WP in 2040 (15 % losses) (kg/m^3)	Potential WP in 2040 (no losses) (kg/m^3)
Nile Delta				
Alexandria	1.93	1.43	1.52	1.90
Demiatte	2.31	1.56	1.45	2.19
Kafr El-Sheik	2.22	1.53	1.45	2.14
El-Dakahlia	2.60	1.37	1.47	2.28
El-Behira	2.37	1.30	1.48	2.16
El-Gharbia	2.42	1.34	1.53	2.21
El-Monofia	2.57	1.38	1.64	2.29
El-Sharkia	2.04	1.08	1.21	1.80
El-Kalubia	1.88	0.99	1.24	1.65
Middle Egypt				
Giza	1.91	1.08	1.42	1.78
Fayoum	1.40	0.88	1.43	1.43
Beni Swief	1.45	0.89	1.39	1.45
El-Minia	1.63	1.01	1.51	1.65
Upper Egypt				
Assuit	1.61	0.84	1.23	1.40
Suhag	1.82	1.05	1.35	1.72
Qena	0.91	0.47	1.13	0.79
Aswan	1.04	0.53	0.71	0.88
Average	1.89	1.10	1.36	1.75

Conclusion

The gap exists between production and consumption of maize can be reduced by improving water management on field level. Our results showed that using surface irrigation involved wasteful use of valuable water resources in Egypt. Thus, changing cultivation method to raised beds and irrigation with drip system can reduce importation by 23 and 12 %, respectively. This situation could occur as a result of increasing productivity per hectare and cultivation of new area with maize. Furthermore, water productivity will increase under these two options compared to its counterpart under surface irrigation.

If we continue to use surface irrigation to cultivate maize under climate change, water requirements will increase, which will reduce maize cultivated area and productivity losses will occur as a result of a biotic stress of climate change. Thus, maize national production will decrease and importation will increase. However, if we implement the above options, we can lower importation to 39 and 22 % under changing cultivation method to raised beds and irrigation with drip system. Furthermore, water productivity will improve compared to its counterpart under surface irrigation.

References

Abou Zeid K (2002) Egypt and the World water goals. Egypt statement in the world summit for sustainable development and beyond, Johannesburg

Abouelenein R, Oweis T, El Sherif M, Awad H, Foaad F, Abd El Hafez S, Hammam A, Karajeh F, Karo M, Linda A (2009) Improving wheat water productivity under different methods of irrigation management and nitrogen fertilizer rates. Egypt J Appl Sci 24(12A):417–431

Chen J, Xu W, Burke JJ, Xin Z (2010) Role of phosphatidic acid in high temperature tolerance in maize. Crop Sci 50:2506–2515

Farré I, Faci JM (2006) Comparative response of maize Zea mays L. and sorghum Sorghum bicolor L. Moench to deficit irrigation in a Mediterranean environment. Agric Water Manag 83:135–143

Heiniger RW (2001) The impact of early drought on corn yield. North Carolina State University. http://www.ces.ncsu.edu/plymouth/cropsci/docs/early_drought_impact_on_corn.html

Hobbs PR, Singh Y, Giri GS, Lauren JG, Duxbury JM (2000) Direct seeding and reduced tillage options in the rice-wheat systems of the Indo-Gangetic plains of South Asia. IRRI workshop, Bangkok, pp 25–26

IPCC Intergovernmental Panel on Climate Change (2007) Intergovernmental panel on climate change fourth assessment report: climate change 2007. Synthesis Report. World Meteorological Organization, Geneva, Switzerland

Karrou M, Oweis T, El Enein R, Sherif M (2012) Yield and water productivity of maize and wheat under deficit and raised bed irrigation practices in Egypt. Afr J Agric Res 711:1755–1760

Lobell DB, Bänziger M, Magorokosho C, Vivek B (2011) Nonlinear heat effects on African maize as evidenced by historical yield trials. Nat Clim Change 1:42–45

Mahmoud I (2014) Simulation of the effect of adaptation strategies on improving yield of some crops grown under climate change conditions, MSc, Institute of Environmental Research, Ain Sham University

Nielsen RL (2007) Assessing effects of drought on corn grain yield. Purdue University, West Lafayette, IN. http://www.kingcorn.org/news/articles.07/Drought-0705.html

Ouda SA, Khalil FA, Yousef H (2009) Using adaptation strategies to increase water use efficiency for maize under climate change conditions. In: 13th international conference on water technology, Hurghada, Egypt. 12–15 March 2009

Ouda S, Abdrabbo M, Noreldin T (2012a) Effect of changing sowing dates and irrigation scheduling on maize yield grown under climate change conditions. In: Proceeding of 4th international conference for field irrigation and agricultural meteorology, Ameria, Egypt, 5–7 Nov 2012

Ouda S, Abdrabbo M, Noreldin T (2012b) Effect of changing sowing dates and irrigation scheduling on maize yield grown under climate change conditions. In: 4th international conference for field irrigation and agricultural metrology, 5–7 Nov 2012

Ouda S, Khalil FA, Noreldin T (2013) Modeling climate change impacts and adaptation strategies for crop production in Egypt: an overview. Springer Publishing House, Berlin

Ouda S, Ibrahim M, Taha A (2014) Water management for maize grown in sandy soil under climate change conditions. Arch Agron Soil Sci. doi:10.1080/03650340.2014.935936

Ouda SA, Noreldin T, Abd El-Latif K (2015) Water requirements for wheat and maize under climate change in North Nile Delta. Span J Agric Res 13(1):1–10, March

Peng S, Huang J, Sheehy JE, Laza RC, Visperas RM, Zhong X, Centeno GS, Khush GS, Cassman KG (2004) Rice yields decline with higher night temperature from global warming. Nat Acad Sci 101:9971–9975

Roeckner E, Bäuml G, Bonaventura L, Brokopf R, Esch M, Giorgetta M, Hagemann S, Kirchner I, Kornblueh L, Manzini E, Rhodin A, Schlese U, Schulzweida U, Tompkins A (2003) The atmospheric general circulation model ECHAM5. Part I: Model description. MPI report 349, Max Planck Institute for Meteorology, Hamburg, Germany, 127 pp

Schussler JR, Westgate ME (1991) Maize kernel set at low water potential: II. Sensitivity to reduce assimilate supply at pollination. Crop Sci 31:1196–1203

Snyder RL, Orang M, Bali K, Eching S (2004) Basic irrigation scheduling BISm. http://www.waterplan.water.ca.gov/landwateruse/wateruse/Ag/CUP/Californi/Climate_Data_010804.xls

Taha A (2012) Effect of climate change on maize and wheat grown under fertigation treatments in newly reclaimed soil. PhD. Thesis, Tanta University, Egypt

Chapter 5
High Water-Consuming Crops Under Control: Case of Rice Crop

Mahmoud A. Mahmoud, Samiha Ouda and Sayed abd El-Hafez

Abstract In this chapter, we calculated the applied water amount for rice in 2013 growing season and investigated the effect of improved cultivation method on rice national production. Moreover, we graphed maximum temperature in 2040 with cutoff temperature to investigate suitability of growing rice in its current growing areas. Under climate change, with the assumption that potential rice production will be reduce by 11 %, its production was assessed in 2020, 2030 and 2040 under traditional and improved cultivation method. The effect of the reduction in the cultivated rice area under climate change was quantified, as well as using potential saved water to cultivate additional area with maize. The results indicated that changing cultivation method from traditional method to cultivation on wide furrows saved a large amount of irrigation water. The saved irrigation amounts were invested to cultivate 309,911 hectare of maize on raised beds, or 371,893 hectare under drip system, which will increase maize production. The results also showed that in 2020, 2030 and 2040 water requirements for rice under traditional planting will increase and consequently its cultivated area will be reduced. The results also showed that during the growing season of rice in both Alexandria and Demiatte, maximum temperature raised above cutoff temperature for a few days during the growing season. Thus, it is implied that Alexandria and Demiatte will be suitable to grow rice in 2040, with probably low yield losses. However, the effect of maximum temperature above cutoff temperature will be more pronounced in the rest of the studied governorates, which might restrict its cultivation in the future. In 2020, 2030 and 2040, water requirements for rice under wide furrows will increase, compared to its values under current climate. Thus, the amount of saved irrigation water, rice productivity and total production will decrease, compared to its counterpart under wide furrows and current climate. Furthermore, the saved water assigned to be use in maize production will be reduced. Water productivity values for rice grown on wide furrows were higher than its value under traditional method in all governorates, either under current climate or climate change.

M.A. Mahmoud (✉) · S. Ouda · S.a. El-Hafez
Water Requirement and Field Irrigation Department, Soils, Water and Environment Research Institute, Agricultural Research Center, Giza, Egypt

© The Author(s) 2016
S. Ouda, *Major Crops and Water Scarcity in Egypt*,
SpringerBriefs in Water Science and Technology,
DOI 10.1007/978-3-319-21771-0_5

Keywords   Rice traditional planting · Rice cultivation on wide furrows · Cutoff temperature · Water productivity

Rice is considered the second important food crop after wheat, and it is a main food crop for more than half of the world population. Rice production plays a significant part in the strategy to overcome food shortage and improve self-sufficiency. It was grown on 573,784 ha in 2013. The average productivity of rice was 7.96 ton/ha. The production of rice in Egypt is higher than its consumption. Thus, it is exported to other countries. Rice is cultivated in eight northern governorates only in the Nile Delta to protect its lands from seawater intrusion. The conventional system for irrigating rice is to flood and maintain free water in the field. Initial flooding provides favorable conditions for land preparation and rapid crop establishment through transplanting and efficient weed management. Continual flooding provides a favorable water and nutrient supply under the anaerobic conditions. However, the conventional system uses a large amount of water because of high water loss through evaporation, seepage, and percolation. A challenge for sustainable rice production is to decrease its applied water and maintain or increase yield through improved water productivity (IRRI 2007).

Previous research on the effect of climate change on rice indicated that yields of rice have been estimated to be reduced by 41 % by the end of the twenty-first century (Ceccarelli et al. 2010). Potential rice yields in China will decrease without CO_2 fertilization effect. With CO_2 effect, the potential yields of rice will increase (Yin et al. 2015). In India, Mishra et al. (2013) concluded that rice yield will decrease by 43 and 25 %, as predicted by two climate change models, i.e., REMO and HadRM3 during 2011–2040. In Korea, significant yield decreases (−22 and −35 %) are expected to occur in 2050 and 2100, respectively, as a result of climate change (Kim et al. 2013). Climate change will aggravate heat, salinity, drought, and submergence stresses for rice plants (Wassmann et al. 2009). In Egypt, Eid (2001) indicated that the Egyptian national rice production is expected to decrease by 11 %, as a result of climate change effects in 2050.

Heat stress has a negative effect on rice growth and yield. Rice is relatively more tolerant to high temperatures during the vegetative phase but highly susceptible during the reproductive phase, particularly at the flowering stage (Jagadish et al. 2010). Temperatures higher than the optimum induced floret sterility and thus decreased rice yield (Nakagawa et al. 2003). High-temperature stress during rice flowering led to decreased pollen production and pollen shed (Prasad et al. 2006). Matsui et al. (2001) indicated that 50 % of spikelet sterility was recorded by 3 °C difference in critical temperature between the tolerant genotypes. Photosynthesis of rice was reduced by 11–36 %, when it was exposed to 3.6 and 7.0 °C higher temperature than ambient from heading to middle ripening stage, respectively (Oh-e et al. 2007). Temperature higher than 35 °C for more than 3 days during repro-ductive stages can affect pollen development and pollination, resulting in decreased seed setting and production. Heat stresses occurred during the reproductive organ developing and flowering stage resulted in severe loss of production (Zou et al. 2009). Temperatures beyond critical thresholds not only reduce the growth duration of the rice crop, but they also increase spikelet sterility, reduce grain-filling duration, and enhance respiratory losses, resulting in lower yield and lower quality of rice grain (Fitzgerald and Resurreccion 2009; Kim et al. 2011). Yoshida (1981) indicated that spikelet sterility in rice occur if temperatures exceed 35 °C at anthesis. Heat stress occurring either during day or night has differential impacts on rice growth and production. High night-time temperatures have been shown to have a greater negative effect on rice yield, with a 1 °C increase above critical temperature (>24 °C) leading to 10 % reduction in both grain yield and biomass (Peng et al. 2004, Welch et al. 2010). The impact of high temperatures at night is more devastating than day-time or mean daily temperatures. Booting and flowering are the stages most sensitive to high temperature, which may sometimes lead to complete sterility. Thus, under climate change, the risk of high-temperature stress on rice productivity would increase notably (Tao and Zhang 2013).

Water shortage is a very important factor on rice production, especially on arid and semi-arid regions, not only for the amount of water deficit but also the time of deficit. Water deficit from booting to grain-filling stage caused greatest rice yield reduction by 77 %, than water stress during which all the growth stages (vegetative, panicle initiation, and boot to grain fill) reduced grain yield and its components (Harbir and Ingram 2000). Water stress tends to delay flowering, and a larger delay in flowering was associated with a higher reduction in grain yield, harvest index, and percentages of fertile panicles and filled grains (Pantuwan et al. 2000). Drought for two weeks from 48 to 62 days after transplanting using withholding irrigation significantly reduced grain yield and grain weight (Ravindra et al. 2002). Rice growth period from booting to grain fill (reproductive stage) was the most sensitive to water deficit (Harbir et al. 2002).

In saturated soil culture system, water was added to raise beds using flood irrigation to form water table that fluctuated between 8 and 15 cm below the bed surface. Saturated soil culture system saved water by up to 32 % compared with permanent flooding (conventional system) according to Tuong et al. (2004) and Humphreys et al. (2004). Direct sown rice in rows or by broadcasting showed lower yields and required 73.5 cm more irrigation water than furrow transplanted rice. Seedlings transplanted in beds and furrows saved approximately 60 cm irrigation water than planting seedlings in flat puddles (Devinder et al. 2005).

Irrigation water saving in rice can be attained by changing cultivation method from basins to wide furrows. Planting in bottom of beds could be applied for the rice cultivars in North Delta Egypt because it increased rice productivity by 3.7 %, enhanced furrows water-use efficiency by 66 %, and saved water by 38 %, compared with traditional planting (Meleha et al. 2008). Naresh et al. (2014) indicated that wide raised beds saved about 15–24 % water and grain yield decreased of about 8 %. Transplanting in the bottom of beds increased productivity of irrigation water by 46 and 33 % compared to transplanting flooded soil and transplanting in bottom of furrows, respectively. Transplanting rice in the bottom of beds saved water by 33 % compared to transplanting under flooding, which normally practiced in North Delta, Egypt (Mahmoud 2012). Aboelenein et al. (2011) stated that growing rice on furrows in Egypt saved 20 % of the applied water and increased rice productivity by 15 %.

Thus, in this chapter, we calculated the applied water amount for rice in 2013 growing season and investigated the effect of improved cultivation method on rice national production. As a result of climate change in 2020, 2030, and 2040, temperature is expected to increase, which will negatively affect growth of rice and might restrict rice growth in its current suitable growing area. Thus, we graphed maximum temperature in 2040 with cutoff temperature. Under climate change, with the assumption that potential rice production will reduce by 11 % (Eid 2001), production was assessed in 2020, 2030, and 2040 under traditional and improved cultivation method. The effect of the reduction in the cultivated rice area under climate change was quantified, as well as the effect on the additional cultivated area with maize.

Water Requirements Under Current Climate and Climate Change

Water requirements for rice under current climate were calculated using BISm (Snyder et al. 2004). Under climate change, A1B climate change scenario was downloaded for each governorate in 2020, 2030, and 2040 using ECHAM5 climate model (Roeckner et al. 2003). The model is an Atmospheric Oceanic General Circulation model, with low resolution, i.e., 1.9×1.9 degree. Water requirements under climate change were calculated using BISm model.

Present Conditions of Rice Production

Table 5.1 presents rice cultivated area, productivity, and total production in 2013 of growing season. Rice cultivated area was 573,784 ha, which produced 5,545,652 ton of grains. About 7,598,433,590 m^3 was used to irrigate that area. This large water amount is a result of low application efficiency under surface irrigation, i.e., 50 %. Furthermore, it is also a result of the traditional method of transplanting rice in flat soil under continuance flooding (pudding).

Potential Rice Yield Grown on Wide Furrows

Results in Table 5.2 showed that if farmers change rice planting method from flat as traditional method to wide furrows as induced method, it will save 1,519,686,718 m^3 for all governorates and total production will also increase to 6,377,500 ton.

Table 5.1 Rice cultivated area, productivity, total production, water requirements per hectare, and total water requirements in the studied governorates

Governorates	Cultivated area (ha)[a]	Productivity (ton/ha)[a]	Total production (ton)[a]	Water requirements (m³/ha)	Total water requirements (m³)
Alexandria	1,059	8.40	8,897	12,540	13,281,950
Demiatte	28,830	8.88	256,014	12,216	352,192,370
Kafr El-Sheik	123,549	9.36	1,156,416	12,372	1,528,545,135
El-Dakahlia	175,675	10.56	1,855,132	13,392	2,352,645,180
El-Behira	87,869	10.08	885,721	13,944	1,225,247,660
El-Gharbia	51,377	9.36	480,886	13,248	680,638,080
El-Sharkia	98,522	8.64	851,231	13,740	1,353,693,425
El-Kalubia	6,903	7.44	51,355	13,356	92,189,790
Total	573,784		5,545,652	13,101	7,598,433,590

[a]Central Administration for Agricultural Economics, 2013

Table 5.2 Water requirements and potential rice production grown on wide furrows and amount of saved water in the studied governorates

Governorates	Water requirements (m³/ha)	Productivity (ton/ha)	Total production (ton)	Amount of saved water (m³)
Alexandria	10032	9.66	10,232	2,656,390
Demiatte	9773	10.21	294,416	70,438,474
Kafr El-Sheik	9898	10.76	1,329,879	305,709,027
El-Dakahlia	10714	12.14	2,133,402	470,529,036
El-Behira	11155	11.59	1,018,579	245,049,532
El-Gharbia	10598	10.76	553,018	136,127,616
El-Sharkia	10992	9.94	978,915	270,738,685
El-Kalubia	10685	8.56	59,058	18,437,958
Total			6,377,500	1,519,686,718

The saved irrigation amounts can be invested in reducing maize production consumption gap. Results in Table 5.3 indicated that the saved irrigation water amounts can cultivate 309,911 ha of maize on raised beds or 371,893 ha under drip system. The production of these cultivated areas will be 3,192,954 and 3,931,499 under raised beds and drip system, respectively.

Table 5.3 Potential maize cultivated area and production using the saved irrigation water amounts

Governorates	Raised beds cultivation		Drip system	
	Cultivated area (ha)	Production (ton)	Cultivated area (ha)	Production (ton)
Alexandria	516	4,584	619	5,645
Demiatte	14,375	145,490	17,250	179,143
Kafr El-Sheik	58,790	606,531	70,548	746,825
El-Dakahlia	100,349	1,095,121	120,419	1,348,428
El-Behira	50,467	520,531	60,561	640,933
El-Gharbia	28,759	294,722	34,511	362,893
El-Sharkia	53,201	494,988	63,842	609,482
El-Kalubia	3,449.93	30,983.96	4,140	38,151
Total	309,911	3,192,954	371,890	3,931,500

Effect of Climate Change on Rice Grown Under Traditional Planting Method

Under climate change condition in 2020, water requirements for rice will increase in all governorates ranging between 6 and 8 %, compared to current water requirements. This increase in water requirements will cause shrinkage in cultivated area by 7 %, compared to current cultivated area. Furthermore, as a result of reduction in rice productivity by 11 %, the total production of rice will decrease in all governorates by 17 %, compared to current total production (Table 5.4).

Table 5.4 Expected rice production under traditional planting method in the studied governorates in 2020

Governorates	Percentage of increase in water requirements (%)	Cultivated area (ha)	Total production (ton)	Percentage of reduction in production (%)
Alexandria	6	998	7,461	16
Demiatte	7	26,877	212,412	17
Kafr El-Sheik	7	115,484	962,027	17
El-Dakahlia	7	163,788	1,539,341	17
El-Behira	8	81,488	731,042	18
El-Gharbia	8	47,704	397,392	17
El-Sharkia	8	91,268	701,818	18
El-Kalubia	8	6,381	42,251	18
Total		533,987	4,593,744	

Table 5.5 Expected rice production under traditional planting method in the studied governorates in 2030

Governorates	Percentage of increase in water requirements (%)	Cultivated area (ha)	Total production (ton)	Percentage of reduction in production (%)
Alexandria	8	978	7,310	18
Demiatte	9	26,561	209,913	18
Kafr El-Sheik	9	112,924	940,706	19
El-Dakahlia	8	163,106	1,532,938	17
El-Behira	9	80,842	725,253	18
El-Gharbia	10	46,721	389,208	19
El-Sharkia	10	89,459	687,904	19
El-Kalubia	12	6,136	40,631	21
Total		526,728	4,533,864	

Under climate change in 2030, results in Table 5.5 showed that the increase in rice water requirements will range between 8 and 12 %, compared to current water requirements. Moreover, the cultivated area and total production will decrease in all governorates by 8 and 18 %, respectively, compared to current climate.

In 2040, rice water requirements will be higher than its counterpart in 2020 and 2030 ranging from 10 to 14 %, while the cultivated area and total production will be lower than its counterpart value in 2020 and 2030, as shown in Table 5.6.

The above results implied that under climate change, rice production will be reduced. Thus, a gap between production and consumption will be created. Taking into consideration population growth, this will widen this food gap.

Table 5.6 Expected rice production under traditional planting method in the studied governorates in 2040

Governorates	Percentage of increase in water requirements (%)	Cultivated area (ha)	Total production (ton)	Percentage of reduction in production (%)
Alexandria	10	963	7,198	19
Demiatte	10	26,209	207,139	19
Kafr El-Sheik	11	111,305	927,217	20
El-Dakahlia	11	158,266	1,487,449	20
El-Behira	11	79,161	710,173	20
El-Gharbia	12	45,872	382,132	21
El-Sharkia	12	87,966	676,424	21
El-Kalubia	14	6,055	43,529	22
Total		515,798	4,441,261	

Effect of Temperature During Rice Growing Season Under Climate Change

The expected rise in the temperature under climate change could cause high stress during rice growing season and lower productivity per hectare to an extent, where it will be not economical to cultivate rice in this area. Thus, we graphed maximum temperature with cutoff temperature for rice (35 °C). We did the comparison in 2040, which represent the worst-case scenario for heat stress.

Figures 5.1 and 5.2 indicate that during the growing season of rice in both Alexandria and Demiatte, maximum temperature raised above 35 °C in few days during the growing season. Thus, it is implied that Alexandria and Demiatte will be suitable to grow rice in 2040, with probably low yield losses.

Regarding to Kafr El-Sheik and El-Dakahlia governorates, Figs. 5.3 and 5.4 indicate that number of days when temperature is higher than 35 °C will be more than its counterpart in Alexandria and Demiatte governorates. These two figures showed that temperature will reach 40 °C or be higher early in the growing season and in mid and late growing season. This result implied that rice productivity will be negatively affected in these two governorates.

Figures 5.5 and 5.6 indicate that number of temperature stress days (above 35 °C) became higher than number of non-stress days (temperature less than 35 °C), which implied that higher yield losses are expected to occur in these two governorates in 2040.

Fig. 5.1 Comparison between maximum temperature and cutoff temperature for rice in Alexandria governorate

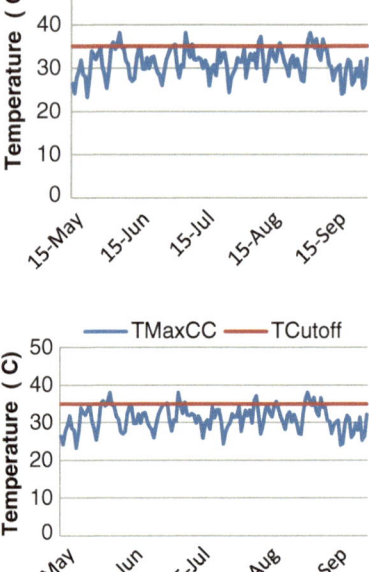

Fig. 5.2 Comparison between maximum temperature and cutoff temperature for rice in Demiatte governorate

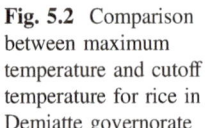

Fig. 5.3 Comparison
between maximum
temperature and cutoff
temperature for rice in Kafr
El-Sheik governorate

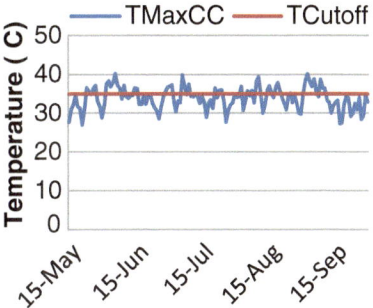

Fig. 5.4 Comparison
between maximum
temperature and cutoff
temperature for rice in
Dakahlia governorate

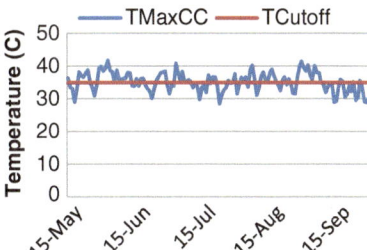

Fig. 5.5 Comparison
between maximum
temperature and cutoff
temperature for rice in
El-Behira governorate

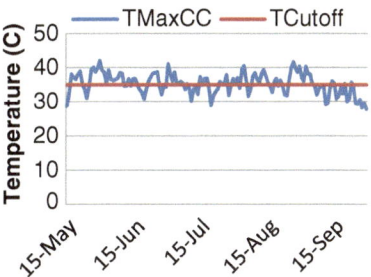

Fig. 5.6 Comparison
between maximum
temperature and cutoff
temperature for rice in
El-Gharbia governorate

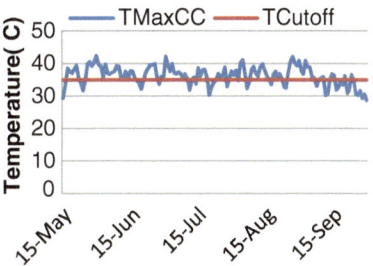

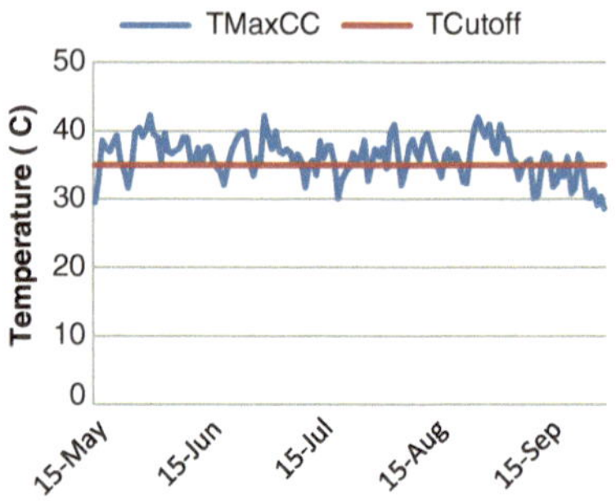

Fig. 5.7 Comparison
between maximum
temperature and cutoff
temperature for rice in
El-Sharkia governorate

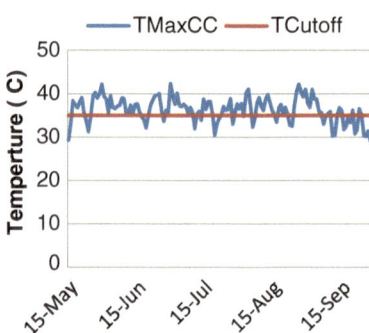

Fig. 5.8 Comparison
between maximum
temperature and cutoff
temperature for rice in
El-Kalubia governorate

Regarding El-Sharkia and El-Kalubia (Figs. 5.7 and 5.8), during rice growing season in 2040, maximum temperature was higher than cutoff temperature during most of the growing season, which will negatively affect rice productivity in these two governorates. Maximum temperature will be higher than 40 °C in many days during whole growing season, and in other days it will reach 45 °C.

Thus, the above results implied that climate change risk on rice production will increase in 2040, as a result of heat stress, which will affect physiological process in the growing plants and results in losses in rice productivity.

Potential Rice Production from Wide Furrows Under Climate Change

In 2020, water requirements for rice under wide furrows will increase for all governorates, compared to its values under current climate. Thus, the amount of saved irrigation water, rice productivity, and total production will decrease, compared to its counterpart under wide furrows and current climate. However, total production of rice under wide furrows in 2020 is expected to be higher than the productivity and total production of rice under traditional method in current climate by 2 % (Table 5.7).

Table 5.7 Water requirements and potential rice production grown on wide furrows and amount of saved water in the studied governorates in 2020

Governorates	Water requirements (m³/ha)	Productivity (ton/ha)	Total production (ton)	Amount of saved water (m³)
Alexandria	10,646	8.59	9,106	2,005,638
Demiatte	10,483	9.09	262,030	49,957,346
Kafr El-Sheik	10,589	9.58	1,183,592	220,312,131
El-Dakahlia	11,491	10.81	1,898,728	333,923,832
El-Behira	12,029	10.32	906,536	168,287,028
El-Gharbia	11,414	9.58	492,186	94,204,256
El-Sharkia	11,866	8.84	871,235	184,669,793
El-Kalubia	11,558	7.62	52,561	12,407,934
Total			5,675,975	1,065,767,958

If we invest the saved irrigation amounts in reducing maize production consumption gap, it can cultivate 192,093 ha on raised beds, or 230,512 ha under drip system. The production of these cultivated areas will be 1,148,773 and 1,414,489 under raised beds and drip system, respectively (Table 5.8).

Under climate change condition in 2030, similar trend will be observed, where water requirements for rice will increase, while the amount of saved water will decrease and the total production is expect to decrease. However, the losses will be higher, compared to the expect values in 2020, as shown in Table 5.9.

Using the saved irrigation amounts maize production will result in cultivating 176,263 ha on raised beds or 211,515 ha under drip system. The production of these cultivated areas will be 1,056,503 and 1,300,877 under raised beds and drip system, respectively (Table 5.10).

Table 5.8 Potential maize cultivated area and production using the saved irrigation water amounts in 2020

Governorates	Raised beds cultivation		Drip system	
	Cultivated area (ha)	Production (ton)	Cultivated area (ha)	Production (ton)
Alexandria	350	1,804	420	2,222
Demiatte	9,229	54,173	11,074	66,703
Kafr El-Sheik	38,204	228,607	45,845	281,485
El-Dakahlia	62,817	397,602	75,380	489,569
El-Behira	30,109	180,119	36,131	221,781
El-Gharbia	17,467	103,819	20,960	127,832
El-Sharkia	31,904	172,164	38,285	211,986
El-Kalubia	2,013	10,485	2,416	12,911
Total	192,093	1,148,773	230,512	1,414,489

Table 5.9 Water requirements and potential rice production grown on wide furrow and amount of saved water in the studied governorates in 2030

Governorates	Water requirements (m³/ha)	Productivity (ton/ha)	Total production (ton)	Amount of saved water (m³)
Alexandria	10867	8.597	9,106	1,771,774
Demiatte	10608	9.089	262,030	46,359,310
Kafr El-Sheik	10829	9.580	1,183,592	190,660,431
El-Dakahlia	11539	10.808	1,898,728	325,491,412
El-Behira	12125	10.317	906,536	159,851,588
El-Gharbia	11654	9.580	492,186	81,873,856
El-Sharkia	12106	8.843	871,235	161,024,493
El-Kalubia	12019	7.615	52,561	9,227,262
Total			5,675,975	976,260,126

Table 5.10 Potential maize cultivated area and production using the saved irrigation water amounts in 2030

Governorates	Raised beds cultivation		Drip system	
	Cultivated area (ha)	Production (ton)	Cultivated area (ha)	Production (ton)
Alexandria	310	1,594	371	1,963
Demiatte	8,564	50,271	10,277	61,899
Kafr El-Sheik	33,063	197,839	39,675	243,600
El-Dakahlia	61,230	387,561	73,476	477,206
El-Behira	28,600	171,090	34,320	210,664
El-Gharbia	15,181	90,230	18,217	111,100
El-Sharkia	27,819	150,120	33,383	184,843
El-Kalubia	1,497	7,798	1,796	9,601
Total	176,263	1,056,503	211,515	1,300,877

The results in Table 5.11 showed that water requirements of rice grown on wide furrows under climate change condition in 2040 will increase by 10–14 %, while the amount of saved irrigation water will decrease, compared to its counterpart under current climate. The total rice productions under wide furrows in 2040 will be higher by 28 % compared with traditional method in 2040, in addition to save an amount of irrigation water equal to 719,873,638 m³.

Results in Table 5.12 indicated that the saved irrigation water amounts can cultivate 129,615 ha on raised beds or 155,538 ha under drip system. Furthermore, the production of these cultivated areas will be 775,011 and 954,274 under raised beds and drip system, respectively.

Table 5.11 Water requirements and potential rice production grown on wide furrows and amount of saved water in the studied governorates in 2040

Governorates	Water requirements (m³/ha)	Productivity (ton/ha)	Total production (ton)	Amount of saved water (m³)
Alexandria	11163	8.597	9,106	1,458,950
Demiatte	10797	9.089	262,030	40,921,399
Kafr El-Sheik	10976	9.580	1,183,592	172,451,246
El-Dakahlia	12146	10.808	1,898,728	218,866,280
El-Behira	12841	10.317	906,536	96,942,672
El-Gharbia	12076	9.580	492,186	60,202,917
El-Sharkia	12503	8.843	871,235	121,903,282
El-Kalubia	12323	7.615	52,561	7,126,891
Total			5,675,975	719,873,638

Table 5.12 Potential maize cultivated area and production using the saved irrigation water amounts in 2040

Governorates	Raised beds cultivation		Drip system	
	Cultivated area (ha)	Production (ton)	Cultivated area (ha)	Production (ton)
Alexandria	255	1,313	306	1,616
Demiatte	7,559	44,375	9,071	54,639
Kafr El-Sheik	29,905	178,944	35,886	220,335
El-Dakahlia	41,172	260,603	49,407	320,882
El-Behira	17,345	103,758	20,813	127,758
El-Gharbia	11,162	66,347	13,395	81,694
El-Sharkia	21,060	113,648	25,272	139,935
El-Kalubia	1,156	6,023	1,387	7,416
Total	129,615	775,011	155,538	954,274

Water Productivity

Water productivity values for rice grown on wide furrows were higher than its value under traditional method in all governorates, either under current climate or climate change (Table 5.13). These results implied that using wide furrows for rice, not only increase productivity per hectare, but also increase water productivity.

Table 5.13 Water productivity (WP) for rice under traditional and wide furrows planting methods in current climate in 2020, 2030, and 2040

Governorates	Current WP (kg/m³)		WP in 2020 (kg/m³)		WP in 2030 (kg/m³)		WP in 2040 (kg/m³)	
	TM	WF	TM	WF	TM	WF	TM	WF
Alexandria	0.67	1.04	0.56	0.81	0.55	0.79	0.54	0.77
Demiatte	0.73	0.96	0.60	0.87	0.60	0.86	0.59	0.84
Kafr El-Sheik	0.76	0.92	0.63	0.90	0.62	0.88	0.61	0.87
El-Dakahlia	0.79	0.88	0.65	0.94	0.65	0.94	0.63	0.89
El-Behira	0.72	0.96	0.60	0.86	0.59	0.85	0.58	0.80
El-Gharbia	0.71	0.98	0.58	0.84	0.57	0.82	0.56	0.79
El-Sharkia	0.63	1.11	0.52	0.75	0.51	0.73	0.50	0.71
El-Kalubia	0.56	1.25	0.46	0.66	0.44	0.63	0.43	0.62
Average	0.69	1.01	0.58	0.83	0.57	0.81	0.56	0.79

Conclusion

Under current condition, planting rice on wide furrows is recommended because it saves a large amount of irrigation water that can be used to grow another crop in the same season like maize to contribute in increasing food security. Wide furrows under current climate have the highest production and highest water productivity. Moreover, under climate change in 2020, 2030, and 2040, wide furrows are recommended for the same reasons.

Developing heat-resistant cultivars to replace heat-sensitive cultivars with adjustment of sowing time and choice of varieties with a short growth duration will allow avoidance of peak stress periods which are some of the adaptive measures that will help in the mitigation of yield reduction due to global warming.

References

Aboelenein R, Sherif M, Karrou M, Oweis T, Benli B, Farahani H (2011) Towards sustainable and improved water productivity in the old lands of Nile Delta. In: Water benchmarks of CWANA —improving water and land productivities in irrigated systems

Ceccarelli S, Grando S, Maatougui M, Michael M, Slash M, Haghparast R, Rahmanian M, Taheri A, Al-Yassin A, Benbelkacem A, Labdi M, Mimoun H, Nachit M (2010) Plant breeding and climate changes. J Agric Sci Camb 148:627–637

Devinder S, Mahey RK, Vashist KK, Mahal SS (2005) Economizing irrigation water use and enhancing water productivity in rice Oryza sativa L. through bed/furrow transplanting. Environ Ecol 23:606–610

Eid HM (2001) Climate change studies on Egyptian agriculture, soils, water and environment research institute SWERI ARC, Ministry of Agriculture, Giza, Egypt

Fitzgerald MA, Resurreccion AP (2009) Maintaining the yield of edible rice in a warming world. Funct Plant Biol 36:1037–1045

FuluT Zhao Z (2013) Climate change, high-temperature stress, rice productivity, and water use in eastern china: a new superensemble-based probabilistic projection. J Appl Meteor Climatol 52:531–551. doi:10.1175/JAMC-D-12-0100.1

Harbir S, Ingram KT (2000) Sensitivity of rice Oryza sativa L. to water deficit at three growth stages. Crop Res Hisar 20:355–359

Harbir S, Ingram KT, Jhorar RK (2002) Comparative performance of different water production functions for rice Oryza sativa L. Crop Res Hisar 23:203–213

Humphreys E, Meisner C, Gupta RK, Timsina J, Beecher HG, Lu TY, Singh Y, Gill MA, Masih Gou ZJ, Hompson JA (2004) Water saving in rice–wheat systems. Paper presented at the 4th international crop science congress in Brisbane, Australia. http://www.cropscience.org.au/icsc2004/poster/1/3/2/852_fujiim.htm. Accessed 5 April 2005

IRRI International Rice Research Institute (2007) Water management in irrigated rice: coping with water scarcity. IRRI, Los Baños Philippines

Jagadish SVK, Muthurajan R, Oane R, Wheeler TR, Heuer S, Bennett J, Craufurd PQ (2010) Physiological and proteomic approaches to dissect reproductive stage heat tolerance in rice Oryza sativa L. J Exp Bot 61:143–156

Kim HY, Ko J, Kang S, Tenhunen J (2013) Impacts of climate change on paddy rice yield in a temperate climate. Glob Change Biol 19:548–562. doi:10.1111/gcb.12047

Kim J, Shon J, Lee CK, Yang W, Yoon Y, Yang WH, Kim YG, Lee BW (2011) Relationship between grain filling duration and leaf senescence of temperate rice under high temperature. Field Crops Res 122:207–213

Mahmoud MA (2012) Effect of deficit irrigation and induced planting methods on water use efficiency, some soil properties and productivity of rice crop in North Nile Delta. Ph.D. thesis, Faculty of Agriculture, Tanta University, Egypt

Matsui T, Omasa K, Horie T (2001) The difference in sterility due to high temperatures during the flowering period among japonica rice varieties. Plant Prod Sci 4:90–93

Meleha ME, El-Bably AZ, Abd Allah AA, El-Khoby WM (2008) Producing more rice with less water by inducing planting methods in north Delta. Egypt J Agric Sci Mansoura Univ 33:805–813

Mishra A, Singh R, Raghuwanshi NS, Chatterjee C, Froebrich J (2013) Spatial variability of climate change impacts on yield of rice and wheat in the Indian Ganga Basin. Sci Total Environ 468–469(Supplement):S132–S138. doi:10.1016/j.scitotenv.2013.05.080

Nakagawa H, Horie T, Matsui T (2003) Effects of climate change on rice production and adaptive technologies. In: Mew TW, Brar DS, Peng S, Dawe D, Hardy B (eds) Rice science: innovations and impact for livelihood. Proceedings of the international rice research conference, Beijing, China, 16–19 Sept 2002, pp. 635–658. Manila, The Philippines: IRRI

Naresh RK, Tomar SS, Kumar Dipender (2014) Experiences with rice grown on permanent raised beds: effect of crop establishment techniques on water use, productivity, profitability and soil physical properties. Rice Sci 21:170–180

Oh-e I, Saitoh K, Kuroda T (2007) Effects of high temperature on growth, yield and dry-matter production of rice grown in the paddy field. Plant Prod Sci 10:412–422

Pantuwan G, Fukai S, Cooper M, Rajatasereekul S, O'Toole JC (2000) Field screening for drought resistance. In: Increased lowland rice production in the Mekong Region: proceedings of an international workshop held in Vientiane, Laos, 30 Oct–2 Nov 2000. ACIAR proceedings no.101, 2001, pp 69–77

Peng SB, Huang JL, Sheehy JE, Laza RC, Visperas RM, Zhong XH, Centeno GS, Khush GS, Cassman KG (2004) Rice yields decline with higher night temperature from global warming. Proc Natl Acad Sci 101:9971–9975

Prasad PVV, Boote KJ, Allen LH, Sheehy JRJE, Thomas JMG (2006) Species, ecotype and cultivar differences in spikelet fertility and harvest index of rice in response to high temperature stress. Field Crops Res 95:398–411

Ravindra K, Tedia K, Malaiya S, Yerne A (2002) Effect of drought on root and shoot growth, plant water status, canopy temperature and yield of rice. J Soils Crops 12:179–182

Roeckner EG, Bäuml Bonaventura L, Brokopf R, Esch M, Giorgetta M, Hagemann S, Kirchner I, Kornblueh L, Manzini E, Rhodin A, Schlese U, Schulzweida U, Tompkins A (2003) The atmospheric general circulation model ECHAM5. Part I: model description. MPI report 349, Max Planck Institute for Meteorology, Hamburg, Germany, 127 pp

Snyder RL, Orang M, Bali K, Eching S (2004) Basic Irrigation Scheduling Bis. http://www.waterplan.water.ca.gov/landwateruse/wateruse/Ag/CUP/Californi/Climate_Data_010804.xls

Tuong TP, Bouman AM, Mortimer M (2004) More rice, less water—integrated approaches for increasing waster productivity in irrigated rice-based systems in Asia. In: 4th international crop science congress on new directions for a diverse planet in Brisbane, Australia. http://www.cropscience.org.au/icsc2004/symposia/1/2/1148_tuongtp.htm.Retrieved01.09

Wassmann R, Jagadish SVK, Heuer S, Ismail A, Redona E, Serra R, Singh RK, Howell G, Pathak H, Sumfleth K (2009) Climate change affecting rice production: the physiological and agronomic basis for possible adaptation strategies. Adv Agron 101:59–122

Welch JR, Vincent JR, Auffhammer M, Moyae PF, Dobermann A, Dawe D (2010) Rice yields in tropical/subtropical Asia exhibit large but opposing sensitivities to minimum and maximum temperatures. Proc Natl Acad Sci 107:14562–14567

Yin Y, Tang Q, Liu X (2015) A multi-model analysis of change in potential yield of major crops in China under climate change. Earth Syst Dynam 6:45–59

Yoshida S (1981) Fundamentals of rice crop science. Los Baños, IRRI, p 269

Zou J, Liu AL, Chen XB, Zhou XY, Gao GF, Wang WF, Zhang XW (2009) Expression analysis of nine rice heat shock protein genes under abiotic stresses and ABA treatment. J Plant Physiol 166:851–861

Chapter 6
High Water Consuming Crops Under Control: Case of Sugarcane Crop

Ahmed M. Taha, Samiha Ouda and Abd El-Hafeez Zohry

Abstract In this chapter, we quantified the effect of the increase in water requirements under climate change on cultivated areas of spring sugarcane. Furthermore, we compared between prevailing temperature during growing season under current climate and cutoff temperature to assess the suitability of these governorates for sugarcane cultivation in 2040. We also investigated the effect using gated pipes to reduce the applied irrigation water to sugarcane. Furthermore, the effect of intercropping summer crops with sugarcane was also investigated to make use of sugarcane applied water and increase water and land productivity. The results indicate that sugarcane production was increased when gated pipes was used for irrigation instead of surface irrigation under both current climate and in 2040. Furthermore, an amount of saved water was attained and could be used in cultivating new land with sugar beet to reduce sugar production-consumption gap in two governorates and in cultivating wheat in the other two governorates. The results also indicate that water requirements for sugarcane will increase by 17 % as an average over all governorates in 2040, which will reduce the cultivated area of sugarcane. Furthermore, comparing measured temperature with predicted temperature in 2040 revealed that it will higher than cutoff temperature from May to September. However, it will be still suitable to grow sugarcane. Intercropping summer oil crops with sugarcane can reduce production-consumption gap in edible oil in Egypt, which will take its water requirements from the applied water to sugarcane. The results also revealed that water and land productivity will increase when gated pipes were used under current and climate change.

Keyword Surface irrigation · Gated pipes · Cutoff temperature · Intercropping · Water and land productivity

A.M. Taha (✉) · S. Ouda
Water Requirement and Field Irrigation Department, Agricultural Research Center,
Soils, Water and Environment Research Institute, Giza, Egypt

A.E.-H. Zohry
Crop Intensification Department, Agricultural Research Center,
Field Crops Research Institute, Giza, Egypt

© The Author(s) 2016   85
S. Ouda, *Major Crops and Water Scarcity in Egypt*,
SpringerBriefs in Water Science and Technology,
DOI 10.1007/978-3-319-21771-0_6

Sugarcane is the main source of sugar in Egypt. It is produced in Upper Egypt only in four governorates, where the prevailing weather conditions are suitable to its production. These governorates are El-Minia, Suhag, Qena, and Aswan. Many industries are dependent on sugarcane cultivation in Upper Egypt. Sugarcane was cultivated on 105,879 ha in 2012/13 with an average productivity equal to 116.6 ton/ha.

Climate change is anticipated to increase temperature. To date, global mean temperatures have increased by about 0.7 °C since the mid-1800s, although the temperature increase is not uniform (IPCC 2007). There are inconsistent results obtained internationally on the effect of climate change on sugarcane, where these researchers concluded that sugarcane yield will be increased or decreased under climate change. Singels et al. (2014) indicated that sugarcane yield will increase under climate change in South Africa, whereas Deressa et al. (2005) reported that an increase by 2 °C will negatively affect sugarcane yield in South Africa. Furthermore, Chandiposha (2013) concluded that sugarcane production will decrease in Zimbabwe. Furthermore, Knox et al. (2010) found a decreasing trend for future projections for sugarcane yield in Swaziland, unless irrigation was included in the simulations. On the other hand, in Brazil, Singels et al. (2014) indicated that the yield will be increased. Furthermore, Marin et al. (2013) reported that in South Brazil, the yield will increase as a result of rain increase in some climate change scenarios or CO_2 fertilization in other scenarios. In Egypt, there were many published papers on the effect of climate change on several crops. However, there was only one published paper on the effect of climate change on sugarcane in Egypt, but we could not obtain it.

However, temperature increase is likely to have an effect on physiological processes of sugarcane plant, since sugarcane is a C4 plant species, whose photosynthetic pathway increases carbon dioxide assimilation with increase in temperature in the range of 8–34 °C (Sage and Kubien 2007). The temperature increase due to climate change is likely to improve sugarcane growth during winter since very low temperatures constrain leaf growth rate and photosynthesis, although this increases sucrose accumulation (Gawander 2007). Furthermore, high temperatures are likely to negatively affect sprouting and emergence of sugarcane (Rasheed et al. 2011). Poor emergence of sugarcane will result in a significantly low plant population. In addition, temperatures above 32 °C results in short internodes, increased number of nodes, and lower sucrose (Bonnett et al. 2006). Thus, climate change is likely to reduce sugarcane and sucrose yields. Furthermore, Clowes and Breakwell (1998) revealed that high temperatures especially at night usually result in more flowering of sugarcane. Flowering in sugarcane ceases growth of leaves and internodes, which reduces sugarcane and sucrose yields. At tillering stage, the crop favored higher minimum temperature (about 26.2 °C), whereas temperatures above 38 °C make sugarcane growth seize (Bonnett et al. 2006).

Furthermore, high daily crop evapotranspiration (ETc) due to high temperatures may cause water stress in sugarcane and that might require applying more frequent irrigation. Since water requirements for sugarcane in Egypt is between 34,255 and 51,055 m^3/ha under surface irrigation with low application efficiency, i.e., 55 %, it is very important to consider ways to reduce these large applied amounts, without any reduction in its productivity. Another way to make use of the high amount of applied water to sugarcane is to use intercropping system, where summer crops, such as soybean, sesame, and sunflower, can be intercropped with spring sugarcane. In this case, these summer crops will use the applied water for sugarcane, without any extra addition of water. Furthermore, under this practice, water and land productivity will increase, as well as farmer's net revenue (Zohry 2005).

In this chapter, we quantified the effect of the increase in water requirements under climate change on cultivated areas of spring sugarcane. Furthermore, we compared between prevailing temperature during growing season of spring sugarcane under current climate and cutoff temperature (above 38 °C) to assess the suitability of these governorates for sugarcane cultivation in 2040. We also investigated the effect using gated pipes to reduce the applied irrigation water to sugarcane. It has been reported by El-Khatib and Sherif (2007) that cultivation ongated pipes increased application efficiency by 10 % and increase productivity by 12 %.

Because water requirements for sugarcane is high, intercropping of summer crops, such as soybean, sesame, and sunflower, with sugarcane was also investigated to make use of sugarcane applied water and increase water and land productivity.

Present Production of Spring Sugarcane

Table 6.1 presents the cultivated area of spring sugarcane, productivity, and total production in 2012/13 growing season. Sugarcane cultivated area was 105,879 ha, which produced 12,438,550 ton. BISm model (Snyder et al. 2004) was used to calculate water requirements per hectare. The total water requirements for sugarcane cultivated area were 4,576,445,141 m^3. This amount of large water is a result of low application efficiency under surface irrigation, i.e. 55 %. Furthermore, high temperature in these four governorates increases evaporation demand and water requirements.

Table 6.1 Sugarcane cultivated area, productivity, total production, water requirements per hectare, and total water requirements in the studied governorates

Governorates	Cultivated area (ha)[a]	Productivity (ton/ha)[a]	Total production (ton)[a]	Water requirements (m^3/ha)[b]	Total water requirements (m^3)
El-Minia	16,456	114.0	1,875,965	34,255	563,687,091
Suhag	6,706	116.9	783,827	36,360	243,839,250
Qena	48,633	120.0	5,835,950	41,716	2,028,788,436
Aswan	34,084	115.7	3,942,808	51,055	1,740,130,364
Total	105,879	116.6	12,438,550	40,846	4,576,445,141

[a]*Source* Central Administration for Agricultural Economics, 2012/13
[b]Calculated with BISm model

Potential Sugarcane Productivity Gates Pipes Under

Irrigation of sugarcane with gated pipes can increase water application efficiency to 70 %. This efficiency can be attributed to short irrigation period which leads to low water percolation into the soil and low evaporation losses from soil surface (El-Berry et al. 2006). Moreover, productivity of sugarcane will be increased by 12 % (El-Khatib and Shrief 2007). Thus, water requirements per hectare will be decreased; productivity and total production will be increased. The total production was increased from 12,438,550 ton (Table 6.1) to 14,926,260 ton (Table 6.2). In addition, an amount of irrigation water could be saved, i.e., 980,666,816 m^3 (Table 6.2).

The amount of saved water in El-Minia and Suhag governorates could be invested in sugar beet cultivation to reduce sugar production–consumption gap. Regarding Suhag and Aswan, sugar beet is not suitable for cultivation there. Therefore, we suggest using these amounts of water to cultivate wheat to reduce its production–consumption gap. Table 6.3 indicates that under surface irrigation, sugar beet-cultivated area could be increased by 17,806 ha and larger area can be cultivated under drip system, i.e., 26,709 ha over the two governorates. Regarding wheat, 112,790 ha can be cultivated under surface irrigation or 150,386 ha can be cultivated under sprinkler irrigation system over the two governorates (Table 6.3).

Table 6.2 Water requirements and potential sugarcane production under gated pipes and amount of saved water in the studied governorates

Governorates	Water requirements (m^3/ha)	Productivity (ton/ha)	Total production (ton)	Amount of saved water (m^3)
El-Minia	26,914	136.80	2,251,158	120,790,091
Suhag	28,569	140.26	940,592	52,251,268
Qena	32,777	144.00	7,003,140	434,740,379
Aswan	40,114	138.82	4,731,370	372,885,078
Total			14,926,260	980,666,816

Table 6.3 Suggested crops to be cultivated in the studied governorates and its cultivated area

Governorates	Suggested crop	Cultivated area under		
		Surface	Sprinkler	Drip
El-Minia	Sugar beet	12,751	–	19,126
Suhag	Sugar beet	5,055	–	7,583
Qena	Wheat	66,373	88,497	–
Aswan	Wheat	46,417	61,890	–

Effect of Climate Change on Sugarcane Grown Under Surface Irrigation

Water requirements for sugarcane will increase by 17 % as an average over all governorates in 2040 under climate change (Table 6.4). With the assumption that no yield losses will occur in 2040, the cultivated area will be reduced to 90,480 ha and the production will be reduced by average of 14 % over the studied governorates (Table 6.4).

The above results are very disturbing because Egypt already has a gap in sugar production–consumption and we need to reduce this gap.

Effect of Temperature Stress on Sugarcane Growing Season

Because we did not find any research on the effect of climate change on sugarcane yield locally, we investigated the effect of temperature stress on sugarcane growing season. It was stated previously that temperatures above 38 °C will seize growth (Bonnett et al. 2006). Thus, we graphed cutoff temperature for sugarcane growth during growing season with measured temperature under current weather and predicted temperature under climate change in each governorate (Fig. 6.1a–d).

In El-Minia and Suhag governorates, measured temperature and predicted temperature in 2040 were higher than cutoff temperature from May to September (Fig. 6.1a, b). However, the number of days where temperature was higher than 38 °C in 2040 was more, compared to the measured temperature values. Similar trends were observed in Qena and Aswan; however, the number of days where temperature is higher than 38 °C is higher and prevailed to October in Aswan (Fig. 6.1c, d). The results in all figures also indicated that the temperature in the studied governorates can sour up 44 °C during the period of May to August in El-Minia and Suhag under climate change in 2040. Furthermore, it could reach 45 and 48 °C in Qena and Aswan governorates, which will cause high temperature stress, especially in Aswan governorate under current climate and under climate change.

Table 6.4 Expected sugarcane production under surface irrigation in the studied governorates in 2040

Governorates	Percentage of increase in water requirements (%)	Cultivated area (ha)	Total production (ton)	Percentage of reduction in production (%)
El-Minia	16	14,177	1,616,216	14
Suhag	16	5,795	677,324	14
Qena	16	41,820	5,018,353	14
Aswan	19	28,688	3,318,657	16
Total		90,480	10,630,550	

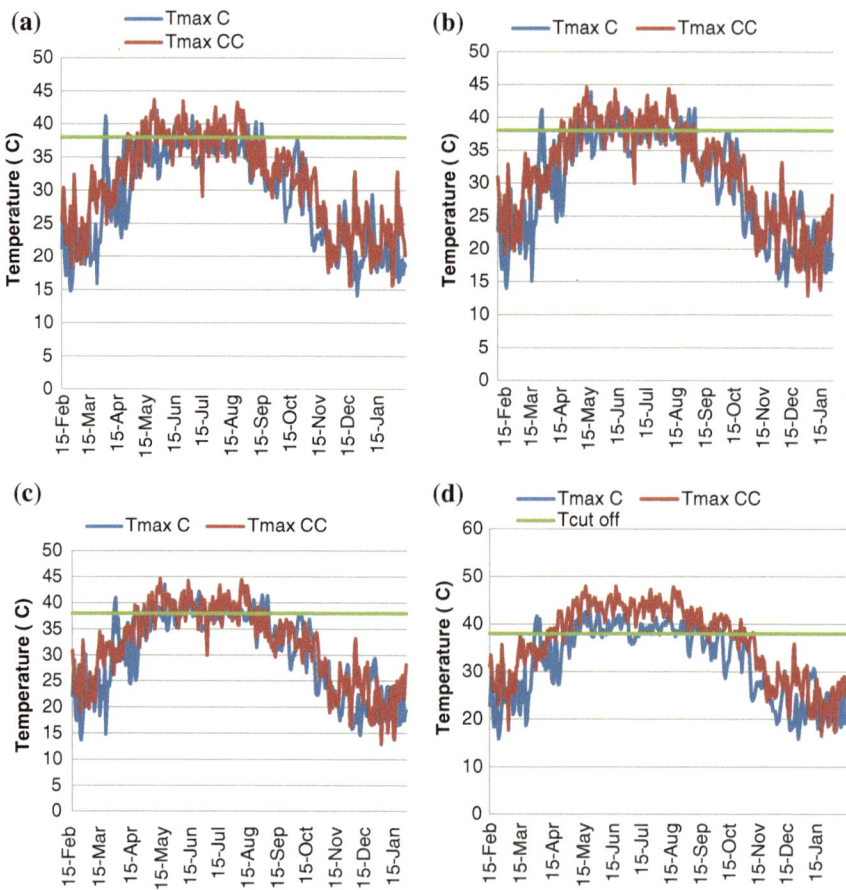

Fig. 6.1 Effect of prevailing temperature (current and climate change) on sugarcane growing season in **a** El-Minia, **b** Suhag, **c** Qena, and **d** Aswan governorates

These findings implied that sugarcane productivity will be reduced in 2040 in this region and it will be still suitable to grow sugarcane. Simulation of sugarcane yield losses under climate change in Egypt was not done. However, for the time being, we assumed that yield losses in El-Minia, Suhage, and Qena are 10 % and in Aswan, we assumed that yield losses will be 15 %.

Potential Sugarcane Yield Under Surface Irrigation in 2040

Extra losses in sugarcane production are expected to occur in 2040 due to the loss in productivity, in addition to loss in cultivated area, where total production will be reduced by 13 % (Table 6.5), compared to reduction due to loss of cultivated area only (Table 6.4).

Governorates	Productivity (ton/ha)	Total production (ton)
El-Minia	102.6	1,454,594
Suhag	105.2	609,592
Qena	108.0	4,516,518
Aswan	98.3	2,820,859
Total		9,401,563

Table 6.5 Expected sugarcane production under surface irrigation in the studied governorates in 2040 (loss in cultivated area and in productivity)

Accordingly, changing surface irrigation to gated pipes in sugarcane can be used as an adaptation strategy to reduce climate change risk on sugarcane production.

Potential Sugarcane Productivity Irrigated with Gated Pipes in 2040

Using gated pipes for irrigation under climate change will result in total production equal to 9,890,984 ton (Table 6.6). As a result of reduction in the applied water using gated pipes, the amount of saved water will be decreased to 362,514,761 m^3 and can be used to cultivate new areas with sugar beet in El-Minia and Suhag and wheat in Qena and Aswan.

Results in Table 6.7 reveal that under surface irrigation, sugar beet-cultivated area could be increased by 7,378 ha and larger area can be cultivated under drip

Table 6.6 Water requirements and potential sugarcane production under gated pipes and amount of saved water in the studied governorates in 2040

Governorates	Water requirements (m^3/ha)	Productivity (ton/ha)	Total production (ton)	Amount of saved water (m^3)
El-Minia	31,240	114.91	1,890,973	49,610,216
Suhag	33,061	117.82	790,097	22,126,008
Qena	38,117	120.96	5,882,638	175,035,920
Aswan	47,659	116.61	3,974,351	115,742,616
Total			9,890,984	362,514,761

Table 6.7 Suggested crops to be cultivated in the studied governorates and its cultivated area

Governorates	Suggested crop	Cultivated area under		
		Surface	Sprinkler	Drip
El-Minia	Sugar beet	5,237	–	7,856
Suhag	Sugar beet	2,141	–	3,211
Qena	Wheat	26,723	35,631	–
Aswan	Wheat	14,408	19,210	–

system, i.e., 11,066 ha over the two governorates. Regarding wheat, 41,131 ha can be cultivated under surface irrigation or 54,841 ha can be cultivated under sprinkler irrigation system over the two governorates.

Intercropping Oil Crops with Spring Sugarcane

An intercropping system is two or more crops sharing the same piece of land for part, or for all, of their growing season (Eskandari et al. 2009). To ensure the optimum productivity in an intercropping system, the peak periods of growth of the two crops should not coincide, so that one quick maturing crop completes its life cycle before the main period of growth of the other crop starts (Parsons 2003). Spring sugarcane (planted in February) offers a unique potential for intercropping. It is planted in wide rows, and takes several months to develop its canopy, during which time the soil and solar energy goes to waste. The growth rate of sugarcane during its early growth stages is slow, with leaf canopy providing sufficient uncovered area for growing of another crop (Nazir et al. 2002).

   Thus, to take advantage of this period, summer oil crops can be intercropped with sugarcane, which will increase land productivity and reduce production–consumption gap in edible oil in Egypt (estimated by 97 %). Furthermore, the intercropped crop will obtaine its water requirements from the applied amount of irrigation water to sugarcane, which will increase water productivity. The intercropping could be done on new cane only. Thus, we suggested that one-third of sugarcane cultivated area could be assigned for intercropping with summer oil crops.

Soybean Intercropping with Spring Sugarcane

Soybean is a very important oil seed and protein crop in the world. The seed contains about 40–45 % protein, 18–20 % edible oil, and 20–26 % carbohydrates. According to Sundara (2000), soybean is one of the important intercrops suitable and compatible with sugarcane. This is mainly due to the fact that soybean has adapted well to the climatic conditions of the sugarcane producing areas and has the greatest potential to fix nitrogen, i.e., up to 300 kg N/ha (Shoko and Tagwira 2005). Since nitrogen fertilizer is a substantial cost component of sugarcane cropping system, the use of soybean as intercropping plays a considerable role in reduction of production costs. El-Geddawy et al. (1988), Zohry (1994), Eweida et al. (1996), and Abou-Kreshe et al. (1997) intercropped soybean with spring sugarcane and they concluded that sugarcane yield was increased, as well as land productivity.

   The total cultivated area of soybean in Egypt was 9,405 ha in 2013 growing season, and its total production was 29,576 ton. Table 6.8 indicates that in 2013 growing season, soybean was not planted in Qena and Aswan governorates. Thus, if we added the assigned sugarcane area to soybean cultivated area, the cultivated

Table 6.8 Potential soybean cultivated area under intercropping with sugarcane in four governorates

Governorates	Sugarcane cultivated area (ha)	Soybean cultivated area (ha)	Total area (sugarcane and soybean area) (ha)
El-Minia	5,485	7,543	13,028
Suhag	2,235	17	2,252
Qena	16,211	–	16,211
Aswan	11,361	–	11,361
Total	35,293	7,560	42,853

soybean area will increase from 7,560 ha in these two governorates to 42,853 ha in four governorates. This increase in the soybean-cultivated area will increase its production and reduce edible oil gap in Egypt.

Sesame Intercropping with Spring Sugarcane

Sesame is an important edible oilseed crop. The seed contains all essential amino acids and fatty acids. It is a good source of vitamins (pantothenic acid and vitamin E) and minerals such as calcium and phosphorous. Furthermore, the seed cake is an important nutritious livestock feed (Balasubramaniyan and Palaniappan 2001). El-Geddawy et al. (1995) and Abou-Keriasha et al. (1997) intercropped sesame with spring sugarcane and indicated that competition over solar radiation between sesame plants and sugarcane plants was low and does not negatively affect sugarcane yield because sesame leaves are erect and do not cause any shading over the growing sugarcane plants.

In Egypt, the total cultivated area of sesame was 17,173 ha in 2013 growing season, and its total production was 198,356 ton. Intercropping sesame with sugarcane can increase its national production by 22,754 ton (Table 6.9).

Table 6.9 Potential sesame cultivated area under intercropping with sugarcane in four governorates

Governorates	Sesame cultivated area (ha)	Total area (sugarcane and sesame area) (ha)	Sesame production from sugarcane area (ton)
El-Minia	2,310	7,795	3,697
Suhag	484	2,719	1,419
Qena	473	16,684	10,651
Aswan	110	11,472	6,987
Total	3,377	38,670	22,754

Sunflower Intercropping with Spring Sugarcane

Sunflower oil is considered as a premium oil because of its light color, mild flavor, low saturated fat levels, and the ability to withstand high cooking temperatures (Weiss 2000). El-Gergawi et al. (2000) indicated that land productivity was increased when sunflower was intercropped with spring sugarcane. However, Abou-Keriasha et al. (1997) indicated that competition over solar radiation between sunflower plants and sugarcane plants was high because sunflower plants were longer than sugarcane plants in that growth stage.

The national production of sunflower was 14,387 ton resulted from 6,025 ha in 2013 growing season. Table 6.10 reveals that cultivated area of sunflower in El-Minia and Suhag governorates was 800 ha. This area can increase by including sugarcane area in the four governorates, which is to be 36,093 ha.

Effect of Changing Irrigation System on Water and Land Productivity

The lowest average water productivity value for sugarcane was obtained using surface irrigation under current climate and under climate change in 2040, i.e., 2.92 and 2.23 kg/m^3, respectively (Table 6.11). Using gated pipes for sugarcane irrigation could increase water productivity, compared to its counterpart under surface irrigation. Similar trend was observed under climate change; however, the value of water productivity under climate change was higher than its counterpart under surface irrigation and current climate (Table 6.11).

Similarly, land productivity follows the same trend as water productivity (Table 6.12). Gated pipes increase land productivity under climate change, compared to its counterpart under surface irrigation and current climate.

These results proved that surface irrigation endures wasteful use of irrigation water, which cannot be tolerated under water scarcity situation.

Table 6.10 Potential sunflower-cultivated area under intercropping with sugarcane in four governorates

Governorates	Sunflower cultivated area (ha)	Total area (sugarcane and sunflower area) (ha)
El-Minia	777	6,262
Suhag	23	2,258
Qena	0	16,211
Aswan	0	11,361
Total	800	36,093

Table 6.11 Water productivity for sugarcane using different irrigation systems under current climate and climate change in 2040

Governorates	Water productivity under current climate (kg/m^3) under		Water productivity under climate change (kg/m^3) under	
	Surface irrigation	Gated pipes	Surface irrigation	Gated pipes
El-Minia	3.33	5.08	2.58	3.68
Suhag	3.21	4.91	2.50	3.56
Qena	2.88	4.39	2.23	3.17
Aswan	2.27	3.46	1.62	2.31
Average	2.92	4.46	2.23	3.18

Table 6.12 Land productivity for sugarcane using different irrigation systems under current climate and climate change in 2040

Governorates	Land productivity under current climate (kg/m^2) under		Land productivity under climate change (kg/m^2) under	
	Surface irrigation	Gated pipes	Surface irrigation	Gated pipes
El-Minia	11.40	13.68	10.26	11.49
Suhag	11.69	14.03	10.52	11.78
Qena	12.00	14.40	10.80	12.10
Aswan	11.57	13.88	9.83	11.66
Average	11.66	14.00	10.35	11.76

Conclusion

Sugarcane is a highly water consuming crop grown in south Egypt under surface irrigation with low application efficiency. There is a gap between sugarcane production and consumption compensated by importation. Changing surface irrigation to gated pipes could increase sugarcane yield and save an amount of irrigation water to be used in cultivation of new land in El-Minia and Suhag with sugar beet and in Qena and Aswan with wheat. Thus, under this system, water and land productivity will increase.

Under climate change in 2040 and under surface irrigation, sugarcane water requirements will increase; as a result its cultivated area will decrease. Furthermore, productivity per hectare will decrease, as well as its national production. Under this disappointing situation, irrigation with gated pipes can reduce yield losses under climate change. Maintain the same cultivated area and still save an amount of irrigation water to expand in new areas with sugar beet and wheat. Thus, water and land productivity will be higher than its counterpart under surface irrigation and current climate. To attain that, the government of Egypt should take care of the costs involved in the construction of gated pipes in sugarcane-cultivated areas to achieve the above benefits.

Another solution to increase water and land productivity in sugarcane-cultivated area is intercropping summer oil crops with sugarcane. Soybean, sesame, and sunflower can be intercropped with sugarcane. These crops will get its water requirements from the applied water to sugarcane. Thus, the cultivated areas of these crops will increase, as well as its national production. This procedure can contribute to reduction of edible oil gap existing in Egypt.

References

Abou-Keriasha M, Zohray AA, Farghly BS (1997) Affect of intercropping some field crops with sugar cane third ratoon. J Agric Sci Mansoure Univ 2212:4163–4176

Balasubramaniyan P, Palaniappan SP (2001) Field crops: an overview. principles and practices of agronomy. Agrobios, India, p 47

Berry CJ, Henson RNA, Shanks DR (2006) On the relationship between repetition priming and recognition memory: Insights from a computational model. J Mem Lang 55:515–533. doi:10.1016/j.jml.2006.08.008

Bonnett GT, Hewitt ML, Glassop D (2006) Effects of high temperature on the growth and composition of sugarcane internodes. Aust J Agric Res 5710:1087–1095

Chandiposha M (2013) Review: potential impact of climate change in sugarcane and mitigation strategies in Zimbabwe. Afr J Agric Res 8:2814–2818

Clowes MJ, Breakwell WL (1998) Zimbabwe sugarcane production manual. Zimbabwe Sugar Association, Chiredzi

Deressa T, Hassan R, Poonyth D (2005) Measuring the economic impact of climate change on South Africa's sugarcane growing regions. Agrekon 44(4):524–542

El-Geddawy IH, Nour AH, Fayed TM, El-Said M (1988) Possibility of intercropping wheat with sugarcane. Commun Sci Dev Res 24(285):110–118

El-Geddawy AS, Laila MS, Nour El-Hoda MT (1995) The relative benefit of intercropping maize with spring cane in Middle Egypt region. Egypt J Appl Sci 10(2):525–632

El-Gergawi ASS, Saif LM, Abou-Salama AM (2000) Evaluation of sunflower intercropping in spring planted sugarcane fields in Egypt. Assuit J Agric Sci 312:163–174

El-Khatib SI, Sherief SA (2007) Effect of surface irrigation systems on yield and yield components of autumn sugarcane and tomato intercropped. Technical conference, 9–11 Dec, San Diego, CA

Eskandari H, Ghanbari A, Javanmard A (2009) Intercropping of cereals and legumes for forage production. Notulae Scientia Biologicae 1:07–13

Eweida MHT, Osman MSA, Shams SAA, Zohry AA (1996) Effect of some intercropping treatments of soybean with sugar cane on growth, yield and quality of both components. Ann Agric Sci, Moshtohor, 24(2):473–486 (Egypt)

Gawander J (2007) Impact of climate change on sugar-cane production in Fiji. WMO Bull 56(1)

IPCC Intergovernmental Panel on Climate Change (2007) Intergovernmental panel on climate change fourth assessment report: climate change 2007. Synthesis report. World Meteorological Organization, Geneva

Knox JW, Rodríguez Díaz JA, Nixon DJ, Mkhwanazi M (2010) A preliminary assessment of climate change impacts on sugarcane in Swaziland. Agric Syst 1032:63–72

Marin F, James R, Jones W, Singels A, Royce F, Assad ED, Pellegrino GQ, Justino F (2013) Climate change impacts on sugarcane attainable yield in southern Brazil. Clim Changes 117:227–239

Nazir MS, Jabbar A, Ahmad I, Nawaz S, Bhatti IH (2002) Production potential and economics of intercropping in autumn-planted sugarcane. Int J Agric Biol 41:140–141

Parsons MJ (2003) Successful intercropping of sugarcane. In: Proceedings of South Africa technology assembly, pp 77–89

Rasheed R, Wahid A, Farooq M, Hussain I, Basra SMA (2011) Role of proline and glycine betaine pretreatments in improving heat tolerance of sprouting sugarcane Saccharum sp. bud. Plant Growth Regul 65:35–45

Sage RF, Kubien DS (2007) The temperature response of C3 and C4 photosynthesis. Plant Cell Environ 30:1086–1106

Shoko MD, Tagwara F (2005) Soybean in sugarcane break crop systems in Zimbabwe: An assessment of potential nutrient and economic benefits. Proc South Afr Sugar Technol Assoc 79:192

Singels A, Jones M, Marin F, Ruane AC, Thorburn P (2014) Predicting climate change impacts on sugarcane production at sites in Australia, Brazil and South Africa using the Canegro model. Sugar Tech 164:347–355

Snyder RL, Orang M, Bali K, Eching S (2004) Basic irrigation scheduling. BIS. http://www.waterplan.water.ca.gov/landwateruse/wateruse/Ag/CUP/Californi/Climate_Data_010804.xls

Sundara B (2000) Sugarcane cultivation. Vikasm Publishing House Private Ltd., India

Weiss EA (2000) Oilseed crops, 2nd edn. Blackwell Science, Oxford

Zohry A.A. 1994. Effect of intercropping soybean with sugarcane. PhD Thesis. Al-Azhar University

Zohry AA (2005) Effect of relaying cotton on some crops under bio-mineral N fertilization rates on yield and yield components. Ann Agric Sci 431:89–103

Chapter 7
Unconventional Solution to Increase Water and Land Productivity Under Water Scarcity

Ahmed Said, Abd El-Hafeez Zohry and Samiha Ouda

Abstract In this chapter we presented an example of prevailing crop rotation in the four soil types exist in Egypt, i.e. old clay, calcareous, sandy and salt affected soils. We also proposed one rotation in each site to replace the prevailing rotation to save on the applied irrigation water. We calculated water requirements for each rotation and determined the amount of saved water per hectare under current climate and in 2040. We also presented the prevailing sugarcane rotation and proposed other rotations to increase water and land productivity. In the proposed rotations, changing cultivation methods from flat or on rows to raised beds saved on the applied water. Furthermore, using intercropping instead of monoculture saved on the applied water under both current climate and in 2040. In the old land, the saved water amounts were 1095, 1331 and 1546 m3/ha in Nile Delta, Middle and Upper Egypt, respectively under current climate. Under climate change, 610, 996 and 1278 m3/ha in Nile Delta, Middle and Upper Egypt, respectively was saved. In the new reclaimed land, the proposed rotation could save 3160 and 2908 m3/ha under current climate and climate change, respectively. Regarding sandy soil, the proposed rotation saved low amount of water, i.e. 53, 67 and 152 m3/ha in Lower, Middle and Upper Egypt, respectively, under current climate. Under climate change, the saved amounts were 58, 66 and 131 m3/ha in Lower, Middle and Upper Egypt, respectively. In the salt-affected soils, the proposed rotation will save 3426 and 2828 m3/ha under current climate and in 2040, respectively. Regarding sugarcane rotations, the amount of saved irrigation water using the proposed rotations was 3,596 and 7,609 m3/ha for spring and autumn rotation, respectively.

Keywords Intercropping · Crop rotations · Clay · Calcareous · Sandy and salt affected soils

A. Said (✉) · A.E.-H. Zohry
Crop Intensification Department, Field Crops Research Institute, Agricultural Research Center, Giza, Egypt

S. Ouda
Water Requirement and Field Irrigation Department, Soils, Water and Environment Research Institute; Agricultural Research Center, Giza, Egypt

© The Author(s) 2016
S. Ouda, *Major Crops and Water Scarcity in Egypt*,
SpringerBriefs in Water Science and Technology,
DOI 10.1007/978-3-319-21771-0_7

Feeding adequately a population growing at an annual rate of 1.84 %, with limited land and water resources, is considered the most important challenge for Egypt. As a result, there is a large gap between production of all strategic crops and its consumption, which increases importation and putting a burden on the country's budget. Agriculture is a vital sector in Egypt's economy, accounting for 14.6 % of GDP. More than 85 % of the water withdrawal from the Nile is used for irrigated agriculture. Water availability, therefore, has a direct influence on national food security. Thus, sustainable growth in agriculture relies on the use of the limited water resources in the most effective and efficient way. At present, surface irrigation was used in over 80 % of Egypt's cultivated land. Poor water management by the Egyptian farmers is contributing to a remarkable waste in irrigation water.

An agricultural management practice that could rationalize the applied irrigation water is the use of crop rotation. Furthermore, it could increase land and water productivity. Crop rotation is one of the most effective agricultural control strategies. It involves arrangement of crops planted on same field and the succeeding crops should belong to different families (Huang et al. 2003). The planned rotation may vary from two to three years, or for a longer period. Some of the general benefits of using rotations are to improve or maintain soil fertility, reduce the spread of pests, reduce risk of weather damage, and increase soil water management, which will be reflected on increasing net profit of farmers. The ultimate goal should be to offer alternatives of different forms of crop rotations with less water requirements and same proportions commodities (cereals, sugar crops, oil crops, and forage crops), as compared to the prevailing crops rotations, which is less benefit to the soil with high water requirements. Kamel et al. (2010) revealed appreciable differences in water consumption between the prevailing rice rotation and proposed rotation. Water consumption of the prevailing exceeded those of the proposed system by 25.1 and 41.7 % in case of the short- and long-term crop rotations, respectively. At present, it is estimated that wheat, clover, cotton, rice, and maize amount for 80 % of the cropped area. Wheat and clover are the principle winter crops. In summer, cotton and rice are important cash crops, while maize and sorghum are major subsistence crops. The inclusion of sugar beet in the crop structure resulted in severe competition with winter crops, particularly faba bean and lentil, which has led to an abrupt decline in the area cultivated by legumes. Intercropping and multiplicity of crop sequence are considered as the successful avenues to increase the cropped area without altering the area cultivated by the main crops in winter or summer. Intercropping is growing two or more crops in the same field, which allow using water and nutrients more efficiently (Eskandari et al. 2009). Andersen (2005) indicated that the advantages of intercropping are as follows: it increases unit land productivity (harvest two types of crops from the same area), increases water productivity (use less water to irrigate two crops), and increases farmers' income (reduce risks from crop failure).

Climate change has the potential to significantly alter the conditions for crop production, with adverse implications on food security in Egypt. Changes in yield behavior in relation to shifts in climate can become critical for the economy of the Egyptian farmers. An increasing probability of low returns as a consequence of the more frequent occurrence of adverse weather conditions could prove dramatic for farmers operating at the limit of economic stress (Torriani et al. 2007). Furthermore, climate change is expected not only to affect crop production, but also it will increase water consumption by crops (Ouda et al. 2015) and that will determine the future of food security in Egypt. Therefore, a more rational use of irrigation must be practiced to conserve irrigation water under current climate condition and to help in fulfilling the anticipated demand under climate change conditions. Since we have to deal with the future and the future by definition is uncertain, we need to be prepared for the worst. Crop rotation could be used as an adaptation strategy to climate change because it consists of crops with medium to low water requirements. In addition, using improved agricultural management practices in cultivating these crops could reduce the applied irrigation water. Previous research in Egypt demonstrated that improved agricultural management practices, such as raised beds cultivation, could save a good percentage of irrigation water, improve the growth environment for the growing crops, and increase the yield of these growing crops, which positively reflected on farmer's net revenue (Abouenein et al. 2009, 2010). Using crop rotations will help in the sustainable use of natural agricultural resources; increase the agricultural productivity of unit land and unit of irrigation water under the prevailing conditions of water scarcity. As a result, the probability of attaining food security for strategic crops will increase and that will help in improving living standards and poverty elevation of the rural population.

In this chapter, we presented an example of prevailing crop rotation in the four soil types existing in Egypt, i.e., old clay soil, calcareous soil, sandy soil, and salt-affected soil. We also proposed one rotation in each site to replace the prevailing rotation to save on the applied irrigation water. We calculated water requirements for each rotation and determined the amount of saved water per hectare under current climate and under climate change in 2040. We also presented the prevailing sugarcane rotation, and we proposed other rotations to increase water and land productivity.

Water Requirements Under Current Climate and Climate Change

Water requirements for crops in each rotation were calculated using BISm (Snyder et al. 2004) under current climate condition. The climate model ECHAM5 (Roeckner et al. 2003) is an Atmospheric Oceanic General Circulation model. The resolution of the model is $1.9 \times 1.9°$. The model was used to develop A1B climate change scenario for each weather station in each governorate in 2040. Water requirements under climate change were also calculated using BISm model.

Crop Rotations in the Old Land

The old irrigated lands are located within the Nile valley and Delta, which represent the most fertile soils in Egypt, are now poor in organic matter content and available nitrogen due to the intensive agricultural system and discontinuation of silt deposits after the construction of the Aswan High Dam. Contentious growing of exhaustive crops like cereals lowered organic matter and reduced microbial activity, which resulted in less ability to hold water and less availability of nutrients in root zones. On the other hand, higher rate of fertilizer to maintain yield leads to NO_3 accumulation in crop root systems. Intensive cropping pattern is existed, where 2 or 3 crops are cultivated each year. After agricultural liberalization in Egypt, cropping pattern has changed gradually towards expanding the production of vegetables at the expense of field crops. Surface irrigation system is prevailing with 60 % application efficiency. There are problems in drainage system and high water table.

An example of the prevailing crop rotation in the old land is presented in Fig. 7.1. It is three-year crop rotation, where the cultivated area is composed of three hectares. Each hectare is divided into three parts and each part is cultivated by winter and summer crops. This rotation can be implemented in all old lands in the Nile Delta and valley. The prevailing rotation is cultivated using the traditional method, i.e., (narrow furrows or flat). It is characterized by high applied irrigation water and high fertilizer consumption. Figure 7.1 shows the prevailing rotation that contained winter legume forage crop (clover), winter and summer cereal crops (wheat and maize), and fiber crop (cotton). The figure also showed that clover is cultivated before maize and before cotton. This practice improves soil fertility and soil sustainability. However, cultivating maize after wheat is exhausting for the soil because both of them are cereal crops.

In the suggested rotation (Fig. 7.2), all crops are cultivated on raised bed to reduce the applied irrigation water and fertilizer. In the first part of the rotation, cotton is relay intercropped with wheat (wheat is cultivated in November and cotton is cultivated in March and harvested in September). The benefits of this system are to increase wheat-cultivated area by the area assigned to be cultivated by cotton and to save the first two irrigations for cotton. Under this system, the farmer obtained two yields: the same cotton yield as if it was planted solely and 80 % of wheat yield compared to sole wheat planting (Zohry 2005b).

Year 1	Year 2	Year 3
Clover	Wheat	Clover (short season)
Maize	Maize	Cotton
Clover (short season)	Clover	Wheat
Cotton	Maize	Maize
Wheat	Clover (short season)	Clover
Maize	Cotton	Maize

Fig. 7.1 Prevailing crop rotation in the old land in Egypt

Year 1	Year 2	Year 3
Wheat/cotton	Sugar beet/faba bean Maize/cowpea forage	Clover Maize/soybean
Clover Maize/soybean	Wheat/cotton	Sugar beet/faba bean Maize/cowpea forage
Sugar beet/faba bean Maize/cowpea forage	Clover Maize/soybean	Wheat/cotton

Fig. 7.2 Proposed crop rotation in the old land in Egypt

In the second part of the rotation, clover is planted as winter crop to increase soil fertility. In the summer season, maize is intercropped with soybean to increase land and water productivity (Ouda et al. 2007). Maize and soybean were intercropped as 2 rows of maize with 2 rows of soybean in one hectare. In this case, maize yield under intercropping was 75 % of its counterpart under sole cultivation. Furthermore, soybean yield under intercropping was 80 % of its counterpart under sole cultivation. This can be attributed to C4 cereal crops, such as maize, which are the dominant plant species, whereas C3 legume crops, such as soybean, are the associated or secondary species. Canopy structures and rooting systems of cereal crops are generally different from those of legume crops. Maize roots can penetrate deeper in the soil (1.7 m) than soybean roots (1.3 m) (Allen et al. 1998). Furthermore, maize can form higher canopy structures than soybean (Ouda et al. 2007).

In the third part of the rotation, faba bean is intercropped with sugar beet as winter crops to reduce faba bean production–consumption gap and increase land and water productivity because no extra irrigation water or fertilizer is applied to faba bean. Sugar beet in cultivated by 100 % of its recommended planting density in one hectare and faba bean is cultivated by 25 % of its recommended planting density in one hectare. As a result, the farmer can obtain 100 and 25 % of sugar beet and faba bean, respectively (Noufal 2012). For summer crops, cowpea is intercropped with maize to increase maize yield and reduce associated weeds. No additional water will be applied for cowpea (Zohry 2005a). Under this system, the farmer use 100 % of plant seeding rate of maize and obtain about 10 % increase in maize yield. Regarding cowpea, the farmer uses 50 % of plant seeding rate of cowpea and obtains 50 % of the yield. Furthermore, it increased farmer's profit (Abou-Keriasha et al. 2011).

Thus, the proposed rotation contains winter and summer cereal crops (wheat and maize) and winter and summer forage crops (clover and cowpea). The rotation also contains winter and summer legume crops (faba bean and soybean), summer fiber crop (cotton), and winter sugar crop (sugar beet). These diversities in the crops that associated in the proposed rotation can serve as a remedy to food insecurity problem existing in Egypt.

Water Requirements for Old Land Rotations Under Current and Climate Change

Under current climate, water requirements for the crops cultivated in the prevailing rotation are 45937, 50262, and 53366 m^3 for three hectares in Nile Delta, Middle Egypt, and Upper Egypt, respectively. Under climate change, water requirements were increased to 51595, 57402, and 61520 m^3 for three hectares (Table 7.1).

Regarding the proposed rotation under current climate, total water requirements for the cultivated crops are 45937, 50262, and 53366 m^3 for three hectares in Nile Delta, Middle Egypt, and Upper Egypt, respectively. Under climate change, total water requirements were increased to 51595, 57402, and 61520 m^3 in Nile Delta, Middle Egypt, and Upper Egypt, respectively. However, if we use the suggested crop rotation, the saved irrigation amount per hectare will be 1095, 1331, and 1546 m^3/ha in Nile Delta, Middle Egypt, and Upper Egypt, respectively, under current climate. Furthermore, the amount of saved water under climate change will be 610, 996, and 1278 m^3/ha in Nile Delta, Middle Egypt, and Upper Egypt,

Table 7.1 Water requirements for crops cultivated in prevailing and proposed rotation in the old land under current climate and climate change

Rotation	Water requirements under current climate (m^3/ha)			Water requirements under climate change (m^3/ha)		
	Nile Delta	Middle Egypt	Upper Egypt	Nile Delta	Middle Egypt	Upper Egypt
Prevailing						
Clover	8530	8850	9080	9383	9912	10260
Maize	6111	7542	7953	6967	8673	9225
Clover (short season)	4265	4425	4540	4692	4956	5130
Cotton	15720	16470	17290	17764	18776	19884
Wheat	5200	5433	6550	5824	6411	7795
Maize	6111	7542	7953	6967	8673	9225
Total	45937	50262	53366	51595	57402	61520
Proposed						
Wheat/cotton	16384	17155	18688	20480	21443	23360
Clover	6824	7080	7264	7506	7930	8208
Maize/soybean	4889	6034	6362	5573	6939	7380
Sugar beet/faba bean	9667	9967	10050	10633	11163	11357
Maize/cowpea	4889	6034	6362	5573	6939	7380
Total	42652	46269	48727	49766	54413	57686
Saved water for 3 ha (m^3)	3285	3994	4639	1829	2989	3834
Saved water (m^3/ha)	1095	1331	1546	610	996	1278

respectively (Table 7.1). These results proved that adopting the proposed crop rotation could reduce the applied water under both current climate and climate change conditions in 2040.

Crop Rotations in the New Reclaimed Land

In new reclaimed lands, calcareous or sandy, soils are deficient in macro- and micro-nutrients. Therefore, there is a need for fertilizer application under intensive agriculture. Nitrogen fertilizer use approximately doubles, while phosphorus application increased. The rate of application of basic nutrients (N, P, and K) is very high. Nitrate leaching to the water table has become a great concern as a potential source of water pollution. Break down of rotation increased concentration of single crop within an area, increased pest infestation, deprived certain elements, and accumulated allelopathy.

In calcareous soil, the problems that threat agricultural sustainability are lack of sustainable rotations, fertility imbalance and ineffective biological N-fixation, soil crust formation, poor irrigation water management and poor drainage, and salinization in some places. Calcareous soil exists in West Delta governorates, such as El-Behira.

An example of a prevailing rotation is presented in Fig. 7.3, where surface irrigation is used in this rotation. Figure 7.3 reveals that the rotation contained winter and summer cereals crops (wheat and maize), a winter forage crop (clover), a vegetable crop (tomato), a legume crop (soybean), and a sugar crop (sugar beet). Thus, the selected crops are suitable to sustain soil fertility; however it does not save irrigation water.

In the suggested rotation (Fig. 7.4), all crops are cultivated on raised beds to save on the applied irrigation water and fertilizer. The proposed rotation includes variety of crop groups. It contains a winter sugar crop (sugar beet), winter and summer legume crops (faba bean and soybean), winter and summer cereal crops (wheat and maize), a vegetable crop (tomato), and winter and summer forage crops (clover and cowpea). In the first part of the rotation, the applied water for faba bean will be saved (Noufal 2012). In the second part of the rotation, wheat will take its water requirements. Intercropping maize with tomato (summer crops) will save the applied irrigation water for maize and tomato yield will increase by 13 %. Regarding maize, 60 % of the

Year 1	Year 2	Year 3
Wheat	Sugar beet	Clover
Tomato	Soybean	Maize
Clover	Wheat	Sugar beet
Maize	Tomato	Soybean
Sugar beet	Clover	Wheat
Soybean	Maize	Tomato

Fig. 7.3 Prevailing crop rotation in calcareous soil in Egypt

Year 1	Year 2	Year 3
Sugar beet/faba bean Maize/soybean	Wheat Tomato/maize	Clover Maize/cowpea forage
Wheat Tomato/maize	Clover Maize/ cowpea forage	Sugar beet/faba bean Maize/soybean
Clover Maize/cowpea forage	Sugar beet/faba bean Maize/soybean	Wheat Tomato/maize

Fig. 7.4 Proposed crop rotation in calcareous soil in Egypt

recommended planting density will be applied and maize yield will be 80 % of its counterpart of sole planting (Abd El-Aal and Zohry 2003). In the third part of the rotation, the required water for clover will be applied. In the summer season, inter-cropping cowpea with maize will save the applied water for cowpea.

Water requirements for prevailing rotation were 47958 and 53684 m^3 under current climate and climate change, respectively, whereas water requirements for the proposed rotation under current climate and climate change will be 38477 and 44961 m^3, respectively. The proposed rotation could save 3160 and 2908 m^3/ha under current climate and climate change, respectively (Table 7.2).

Regarding sandy soil, many constrains exist to maintain sustainability, such as, low soil fertility, ineffective biological N-fixation, poor irrigation water manage-ment, high water table levels, and soil salinity in some areas. Sprinkler or drip system prevailed in this region, with application efficiency 80 and 95 %, respectively.

Figure 7.5 shows the prevailed crop rotation in sandy soil. There are a winter forage crop (clover), winter and summer cereal crops (wheat and maize), a sugar crop (sugar beet), and a winter legume crop (peanut) included in the rotation. All these crops are suitable and maintain soil fertility, except sugar beet followed by maize and wheat followed by peanut.

The proposed crop rotation (Fig. 7.6) indicated that sunflower can intercrop with soybean, where the recommended intercropping pattern should be 2:2, respectively (El-Yamani et al. 2010). Sesame can intercrop with peanut saved, where all the applied water to sesame can be saved. In this system, peanut planting density is 100 %, compared to its sole planting and intercropped sesame density is 25 % from its recommended planting density (Abou-Kerisha et al. 2008). Thus, the proposed rotation included winter and summer forage and cereal crops. It also contains oil summer crops (sunflower and sesame) and sugar crop.

In sandy soil, the crops grown in both rotations, prevailed and proposed, were irrigated with sprinkler or drip system. The proposed rotation saved low amount of water, i.e., 53, 67, and 152 m^3/ha in Lower, Middle, and Upper Egypt, respectively, under current climate, whereas under climate change the saved amounts were lower compared to its counterpart under current climate, i.e., 58, 66, and 131 m^3/ha in Lower, Middle, and Upper Egypt, respectively. The reduction in the amount of applied water is due to replacing maize with sunflower intercropped with soybean, which has lower water requirements than maize (Table 7.3).

Table 7.2 Water requirements for crops cultivated in prevailing and proposed rotation in calcareous land under current climate and climate change

Rotation	Water requirements (m³/ha) under	
	Current climate	Climate change
Prevailing		
Wheat	5017	5368
Tomato	13170	15145
Clover	8210	8949
Maize	6069	6980
Sugar beet	9561	10422
Soybean	5931	6821
Total	47958	53684
Proposed		
Sugar beet/faba bean	7649	8796
Maize/soybean	4856	5730
Wheat	4013	4615
Tomato/maize	10536	12537
Clover	6568	7553
Maize/cowpea	4856	5730
Total	38477	44961
Saved water for 3 ha (m³)	9481	8723
Saved water (m³/ha)	3160	2908

Year 1	Year 2	Year 3
Clover Maize	Wheat Peanut	Sugar beet Soybean
Sugar beet Maize	Clover Maize	Wheat Peanut
Wheat Peanut	Sugar beet Maize	Clover Maize

Fig. 7.5 Prevailed crop rotation in sandy soil in Egypt

Year 1	Year 2	Year 3
Clover Maize/cowpea forage	Wheat Peanut/sesame	Sugar beet/faba bean Maize/potato
Sugar beet/faba bean Sunflower/soybean	Clover Maize/cowpea forage	Wheat Peanut/sesame
Wheat Sesame/peanut	Sugar beet/faba bean Maize/potato	Clover Maize/cowpea forage

Fig. 7.6 Proposed crop rotation in sandy soil in Egypt

Table 7.3 Water requirements for crops cultivated in prevailing and proposed rotation in sandy soil under current climate and climate change

Rotation	Water requirements under current climate (m³/ha)			Water requirements under climate change (m³/ha)		
	Lower Egypt	Middle Egypt	Upper Egypt	Lower Egypt	Middle Egypt	Upper Egypt
Prevailing						
Clover	8213	8500	8750	8787	9435	9713
Maize	5002	5740	6381	5007	6314	7019
Sugar beet	7521	8273	9101	8048	9184	10102
Maize	5002	5377	6131	5502	5915	6744
Wheat	4850	5335	5869	5335	5869	6455
Peanut	5802	6052	6252	6382	6574	6771
Total	36390	39277	42484	39062	43289	46804
Proposed						
Clover	8213	8500	8750	8787	9435	9713
Maize/cowpea	5002	5740	6381	5007	6314	7019
Sugar beet/faba bean	7521	8273	9101	8048	9184	10102
Sunflower/soybean	4843	5176	5676	5327	5694	6244
Wheat	4850	5335	5869	5335	5869	6455
Sesame/peanut	5802	6052	6252	6382	6597	6877
Total	36231	39076	42029	38887	43091	46410
Saved water for 3 ha (m³)	159	201	455	175	198	394
Saved water (m³/ha)	53	67	152	58	66	131

Crop Rotations in Salt-Affected Soils

The majority of salt-affected soils in Egypt are located in the Northern Nile Delta. About 900,000 hectares of Egypt's agricultural lands are suffering from salinity build-up problem. Across Egypt's agricultural land profile, salt-affected soils represent about 60 and 25 % in North and South Nile Delta regions, respectively. Surface irrigation system with either fresh water, agricultural drainage or mixed water is prevailing. There are several problems here, such as high water table, seawater intrusion, and increase in soil salinity.

An example of prevailing rotation is illustrated in Fig. 7.7. The rotation is characterized by intensive rice cultivation to leach salts away from root zone. Furthermore, the rotation contains forage, cereal, and sugar crops. However, maize cultivation after wheat and rice after sugar beet are exhausting to the soil.

Cultivation on raised beds is the main characteristic of the proposed rotation (Fig. 7.8). Cultivation of short season clover after rice and before wheat increased

Year 1	Year 2	Year 3
Clover	Sugar beet	Wheat
Maize	Rice	Rice
Wheat	Clover	Sugar beet
Rice	Maize	Rice
Sugar beet	Wheat	Clover
Rice	Rice	Maize

Fig. 7.7 Prevailing crop rotation in salt-affected soil in Egypt

Year 1	Year 2	Year 3
Wheat Rice Clover (short season)	Sugar beet/faba been Maize/cowpea forage	Wheat/cotton
Sugar beet/faba been Maize/cowpea forage	Wheat/cotton	Wheat Rice Clover (short season)
Wheat/cotton	Wheat Rice Clover (short season)	Sugar beet/faba been Maize/cowpea forage

Fig. 7.8 Proposed crop rotation in salt-affected soil in Egypt

soil fertility, compared to wheat and rice cultivation only (Sheha et al. 2015). Furthermore, the rotation contains winter and summer cereal crops and winter and summer forage crops. Furthermore, it contains sugar crop and fiber crop.

Cultivating crops in salt-affected soil requires application of leaching requirements to wash salts away from root zone. Table 7.4 includes water requirements for both prevailing and proposed crop rotations without leaching requirements. The proposed rotation will save 3426 and 2828 m^3/ha under current climate and under climate change conditions in 2040, respectively.

Sugarcane Rotations in Upper Egypt

Sugarcane crop is considered very exhaustive to the soil, and requires great quantities of irrigation water, fertilizers, and labor force. It is a perennial crop that usually stays in field for three years (two ratoons, in addition to the yield of new cane). Thereafter, the soil requires more special sustainable rotation to conserve soil fertility. Sugarcane belt is located in the Southern governorates, where 90 % is grown in El-Minia, Suhag, Qena, and Aswan governorates.

Usually, fields are left fallow after latter ratoon during winter, and then, farmers grow grain sorghum. Wheat and Egyptian clover are usually grown in sugarcane rotation (Zohry 1994).

Table 7.4 Water requirements for crops cultivated in prevailing and proposed rotation in salt-affected land under current climate and climate change

Rotation	Water requirements (m³/ha) under	
	Current climate	Climate change
Prevailing		
Clover	8720	9592
Maize	6361	7252
Wheat	5600	6608
Rice	13740	15389
Sugar beet	9667	10634
Rice	13740	15389
Total	57828	64863
Proposed		
Wheat	4480	5617
Rice	10992	12311
Clover (short season)	2790	3069
Sugar beet/faba bean	7734	8507
Maize/cowpea forage	5089	5801
Wheat/cotton	16464	21074
Total	47549	56379
Saved water for 3 ha (m³)	10279	8483
Saved water (m³/ha)	3426	2828

Prevailing Crop Rotation for Sugarcane

Farmers follow this rotation in large area and in most fertile soil in Upper Egypt. Nevertheless, this rotation suffers irregularities in income and work organization and the scarcity of seeds (stakes) to plant the fourth year. In this rotation, sugarcane occupies half the area, stays in field for 3 years, and renews every year. Cane is grown in autumn, and the Egyptian clover of one cut as a cover crop is always ploughed in the soil before planting cane. The field is divided in to six fields and this necessitates the design of preliminary years before starting the rotation. Furthermore, cane occupies half the area. This rotation is characterized by cultivation of wheat, clover, and legumes in the winter and grain sorghum in the summer (Fig. 7.9).

The proposed autumn sugarcane rotation (Fig. 7.10) is characterized by using intercropping with sugarcane, which allows saving of the applied water to the intercropped crop as it uses the applied water to sugarcane (Zohry 2005a). Thus, intercropping faba bean or onion with autumn planted sugarcane can be done. Under that system, intercropping two rows of faba bean between sugarcane ridges or five rows of onion with sugarcane was successful and profitable (Farghly 1997; Zohry 1997). Furthermore, cowpea is intercropped with grain sorghum to improve soil fertility, as cowpea is a legume crop (Abou-Keriasha et al. 2011). Intercropping

Plot area	First preliminary	Second Preliminary	First year	Second year	Third year	Fourth year	Fifth year	Sixth year
1/6	Egyptian clover (one cut) New cane	1st Ratoon	2nd Ratoon	Fallow Grain sorghum	Wheat Grain sorghum	Legume Grain sorghum	Egyptian clover (one cut) New cane	1st Ratoon
1/6	Legume Grain sorghum	Egyptian clover (one cut) New cane	1st Ratoon	2nd Ratoon	Fallow Grain sorghum	Wheat Grain sorghum	Legume Grain sorghum	Egyptian clover (one cut) New cane
1/6	Wheat Grain sorghum	Legume Grain sorghum	Egyptian clover (one cut) New cane	1st Ratoon	2nd Ratoon	Fallow Grain sorghum	Wheat Grain sorghum	Legume Grain sorghum
1/6	Legume Grain sorghum	Wheat Grain sorghum	Legume Grain sorghum	Egyptian clover (one cut) New cane	1st Ratoon	2nd Ratoon	Fallow Grain sorghum	Wheat Grain sorghum
1/6	Wheat Grain sorghum	Legume Grain sorghum	Wheat Grain sorghum	Legume Grain sorghum	Egyptian clover (one cut) New cane	1st Ratoon	2nd Ratoon	Fallow Grain sorghum
1/6	Legume Grain sorghum	Wheat Grain sorghum	Legume Grain sorghum	Wheat Grain sorghum	Legume Grain sorghum	Egyptian clover (one cut) New cane	1st Ratoon	2nd Ratoon

Fig. 7.9 Prevailing sugarcane rotation in Upper Egypt

maize and soybean could replace grain sorghum, which use similar irrigation water. Sunflower is also cultivated in this rotation, as well as maize. Thus, this rotation consumes similar amount of irrigation water as prevailing rotation. However, intercropping increases water and land productivity as more than one crop consumed the same amount of applied water and more than one crop were harvested from the same piece of land area.

The proposed spring sugarcane rotation (Fig. 7.11) is characterized by intercropping summer crops with sugarcane, where these crops will obtain their water requirements from the applied water to sugarcane. Intercropping soybean or sesame with spring planted sugarcane, where two rows of soybean or sesame will be planted between sugarcane ridges, was successful and profitable (Eweida et al. 1996; Abou-Keriasha et al. 1997). Similar to autumn sugarcane rotation, summer crops, such as sunflower, soybean, and maize, could replace grain sorghum in some areas in the rotation. Furthermore, intercropping maize and cowpea, as well as maize and soybean could be implemented.

Plot area	First preliminary	Second Preliminary	First year	Second year	Third year	Fourth year	Fifth year	Sixth year
$1/6$	Egyptian clover (one cut) New cane + Faba bean	1st Ratoon	2nd Ratoon	Faba bean Grain sorghum	Wheat Soybean + Maize	Faba bean Grain sorghum	Egyptian clover (one cut) New cane + Faba bean	1st Ratoon
$1/6$	Faba bean Grain sorghum	Egyptian clover (one cut) New cane + Onion	1st Ratoon	2nd Ratoon	Faba bean Grain sorghum	Wheat Maize + Soybean	Faba bean Grain sorghum	Egyptian clover (one cut) New cane
$1/6$	Wheat Grain sorghum + Cowpea	Faba bean Grain sorghum + Cowpea	Egyptian clover (one cut) New cane + Faba bean	1st Ratoon	2nd Ratoon	Clover Grain sorghum	Wheat Soybean	Faba bean Grain sorghum
$1/6$	Clover Grain sorghum	Wheat Sunflower	Faba bean Maize + Soybean	Egyptian clover (one cut) New cane + Onion	1st Ratoon	2nd Ratoon	Clover Grain sorghum	Wheat Maize + Soybean
$1/6$	Wheat Sunflower	Clover Grain sorghum	Wheat Maize	Clover Maize + Cowpea	Egyptian clover (one cut) New cane + Faba bean	1st Ratoon	2nd Ratoon	Clover Sunflower
$1/6$	Faba bean Grain sorghum	Wheat Grain sorghum + Cowpea	Clover Maize + Soybean	Wheat Grain sorghum	Clover Maize	Egyptian clover (one cut) New cane + Onion	1st Ratoon	2nd Ratoon

Fig. 7.10 Proposed autumn sugarcane rotation in Upper Egypt

Amount of Saved Irrigation Water Under Proposed Rotations

The amount of saved irrigation water using the proposed spring rotation was calculated to be 3,596 m^3/ha (Table 7.5). This amount resulted from intercropping soybean or sesame with sugarcane. This amount of water was supposed to be applied to either crops under sole planting. Thus, it was saved because either crops obtained it from the applied water to sugarcane. Similarly, under proposed autumn rotation, the saved amount was 7,609 m^3, as a result of intercropping faba bean or onions (Table 7.5).

Plot area	First preliminary	Second Preliminary	First year	Second year	Third year	Fourth year	Fifth year	Sixth year
$1/6$	Egyptian clover (one cut) New cane + Soybean	1st Ratoon	2nd Ratoon	Faba bean Grain sorghum	Wheat Soybean + Maize	Faba bean Grain sorghum	Egyptian clover (one cut) New cane + Soybean	1st Ratoon
$1/6$	Wheat Grain sorghum	Egyptian clover (one cut) New cane + Sesame	1st Ratoon	2nd Ratoon	Faba bean Grain sorghum	Wheat Maize + Soybean	Faba bean Grain sorghum	Egyptian clover (one cut) New cane + Sesame
$1/6$	Faba bean Grain sorghum + Cowpea	Faba bean Grain sorghum	Egyptian clover (one cut) New cane + Soybean	1st Ratoon	2nd Ratoon	Clover Grain sorghum	Wheat Soybean	Faba bean Grain sorghum
$1/6$	Clover Grain sorghum	Wheat Sunflower	Faba bean Maize + Soybean	Egyptian clover (one cut) New cane + Sesame	1st Ratoon	2nd Ratoon	Clover Grain sorghum	Wheat Maize + Soybean
$1/6$	Wheat Sunflower	Clover Grain sorghum	Wheat Maize	Clover Maize + Cowpea	Egyptian clover (one cut) New cane + Soybean	1st Ratoon	2nd Ratoon	Clover Sunflower
$1/6$	Faba bean Grain sorghum	Wheat Grain sorghum + Cowpea	Clover Soybean	Wheat Grain sorghum	Clover Maize	Egyptian clover (one cut) New cane + Sesame	1st Ratoon	2nd Ratoon

Fig. 7.11 Proposed spring sugarcane rotation in Upper Egypt

Table 7.5 Saved amount of irrigation water under proposed rotations

	Prevailing rotation		Proposed rotation	
	Spring	Autumn	Spring	Autumn
Applied water per hectare (m³/ha)	19,708	20,958	19,708	20,958
Amount of saved water (m³/ha)	–	–	3,596	7,609

Conclusion

Climate change is expected to negatively affect water resources in Egypt and worsen water scarcity situation that already existed. Thus, we are forced to think about unconventional procedures to increase crops production, maintain irrigation water, and conserve irrigation water. These procedures are implemented easily by farmers, do not involve any extra cost, and result in an increase in farmer's net revenue. Crop rotation and intercropping can help in solving food insecurity problem through increase in land productivity. Furthermore, implementing crop rotation and intercropping can save sum of irrigation water and increase water productivity. These saved irrigation water amounts can be used to cultivate new areas and reduce food insecurity under current climate. Under climate change in 2040, these saved amounts can be used to maintain the cultivated area that is expected to reduce, as a result of the increase in water requirements of the cultivated crops.

The proposed sugarcane (autumn or spring) rotations can maintain soil fertility (legumes versus cereals), increase water productivity (an amount of water irrigate two crops), and increase land productivity (two crops harvested from the same area). As a result, farmer's income can increase and livelihood of the farmers can be improved.

References

Abd El-Aal AIN, Zohry AA (2003) Natural phosphate affecting maize as a protective crop for tomato under environmental stress conditions at Toshky. Egypt J Agric Res 81 3:937–953 (Egypt)

Abou-Keriasha MA, Gadallah RA, Mohamdain EEA (2008) Response of groundnut to intercropping with some sesame varieties under different plant density. Arab Univ J Agric Sci Ain Shams Univ Cairo 16(2):359–375 (Egypt)

Abou-Keriasha MA, Zohry AA, Farghly BS (1997) Effect of intercropping some field crops with sugar cane on yield and its components of plant cane and third ratoon. J Agric Sci Mansoura Univ 22(12):4163–4176 (Egypt)

Abouelenein R, Oweis T, El Sherif M, Awad H, Foaad F, Abd El Hafez S, Hammam A, Karajeh F, Karo M, Linda A (2009) Improving wheat water productivity under different methods of irrigation management and nitrogen fertilizer rates. Egypt J Appl Sci 24(12A):417–431 (Egypt)

Abouenein R, Oweis T, Sherif M, Khalil FA, Abed El-Hafez SA, Karajeh F (2010) A new water saving and yield increase method for growing berseem on raised seed bed in Egypt. Egypt J Appl Sci 25(2A):26-41 (Egypt)

Abou-Keriasha MA, Ibrahim TS, Mohammadain EE (2011) Effect of Cowpea intercropping date in Maize and Sorghum fields on productivity and infestation weed. Egypt J Agron 33(1):35–49

Allen RG, Pereira LS, Raes D, Smith M (1998) Crop evapotranspiration: guideline for computing crop water requirements. FAO, No56

Andersen MK (2005) Competition and complementarily in annual intercrops—the role of plant available nutrients. Ph.D. thesis, Department of Soil Science, Royal Veterinary and Agricultural University, Copenhagen, Denmark. Samfundslitteraur Grafik, Frederiksberg, Copenhagen

El-Yamani KH, Ibrahim Sahar T, Osman EBA (2010) Intercropping sunflower with soybean in the Newly Valley. Egypt J Appl Sci 25(7):288–302 (Egypt)

Eskandari H, Ghanbari A, Javanmard A (2009) Intercropping of cereals and legumes for forage production. Not Sci Biol 1(1):7–13

Eweida MHT, Osman MSA, Shams SAA, Zohry AA (1996) Effect of some intercropping treatments of soybean with sugar cane on growth, yield and quality of both components. Ann Agric Sci Moshtohor 24(2):473–486 (Egypt)

Farghly BS (1997) Yield of sugar cane as affected by intercropping with faba bean. J Agric Sci Mansoura Univ 22(12):4177–4186 (Egypt)

Huang M, Shao M, Zhang L, Li Y (2003) Water use efficiency and sustainability of different long-term crop rotation systems in the Loess Plateau of China. Soil Tillage Res 72:95–104. doi:10.1016/s0167-19870300065-5

Kamel AS, El-Masry ME, Khalil HE (2010) Productive sustainable rice based rotations in saline-sodic soils in Egypt. Egypt J Agron 32(1):73–88

Nofal N (2012) Effect of intercropping faba bean on sugar beet under different nitrogen fertilization. M.Sc. thesis, Faculty of Agric., El-Minia University, Egypt

Ouda SA, Noreldin T, Abd El-Latif K (2015) Water requirements for wheat and maize under climate change in North Nile Delta. Span J Agric Res 13(1):e03–001

Ouda SA, El Mesiry T, Abdallah EF, Gaballah MS (2007) Effect of water stress on the yield of soybean and maize grown under different intercropping patterns. Aust J Basic Appl Sci 14:578–585

Roeckner E, Bäuml G, Bonaventura L, Brokopf R, Esch M, Giorgetta M, Hagemann S, Kirchner I, Kornblueh L, Manzini E, Rhodin A, Schlese U, Schulzweida U, Tompkins A (2003) The atmospheric general circulation model ECHAM5. Part I: model description. MPI report 349, Max Planck Institute for Meteorology, Hamburg, Germany, 127 pp

Sheha MA, Ahmed RN, Abou-Elela MA (2015) Effect of crop sequence and nitrogen levels on rice productivity. Under publication

Snyder RL, Orang M, Bali K, Eching S (2004) Basic irrigation scheduling (BIS). http://www.waterplan.water.ca.gov/landwateruse/wateruse/Ag/CUP/Californi/Climate_Data_010804.xls

Torriani DS, Calanca P, Schmid S, Beniston M, Fuhrer J (2007) Potential effects of changes in mean climate and climate variability on the yield of winter and spring crops in Switzerland. Clim Res 34:59–69

Zohry AA (1994) Effect of intercropping soybean with sugarcane. Ph.D. thesis, Al-Azhar University

Zohry AA (1997) Effect of intercropping onion with autumn planted sugarcane on can yield and juice quality. Egypt J Agric Res 771:273–287

Zohry AA (2005a) Effect of preceding winter crops and intercropping on yield, yield components and associated weeds in maize. Ann Agric Sci Moshtohor 43(1):139–148

Zohry AA (2005b) Effect of relaying cotton on some crops under bio-mineral N fertilization rates on yield and yield components. Ann Agric Sci 431:89–103

Chapter 8
Recommendations to Policy Makers to Face Water Scarcity

Sayed A. Abd El-Hafez and A.Z. El-Bably

Abstract This chapter provides insights to policy makers in Egypt on how to deal with water scarcity under current climate and in 2040 under climate change.

Policy response to climate variability and change should be flexible and sensible. The difficulty of prediction and the impossibility of verification of predictions decades into the future are important factors that allow for competing views of long-term climate future. Therefore, policies related to long-term climate should not be based on particular predictions, but instead should focus on policy alternatives that make sense for a wide range of plausible climatic conditions. Climate is always changing on a variety of time scales, and being prepared for the consequences of this variability is a wise policy.

Climate change will have far-reaching effects on water management in agriculture, even if adaptive capacity is relatively strong. In Egypt, the impacts will vary considerably from location to location, but will arise through a combination of less favorable conditions for plant growth, such as more variable rainfall, lower water availability for irrigation, and higher crop water demands. These stresses will be additional to the pressures to produce more food, with less water and less land degradation in the face of the rising of the population and changing food preferences.

In response to climate change, estimates of incremental water requirement to meet the future demand in 2040 to agricultural production under climate change vary from 5 to 10 % of the extra water needed. It ranges from 2 to 19 % for wheat, 11 to 19 % for maize, 10 to 14 % for rice, and 11 to 19 % for sugarcane. One consequence of greater future water demand and likely reductions in supply is that the emerging competition between the environment and agriculture for water will be much greater, and the matching of supply and demand consequently will be harder to reconcile. The future availability of water to match crop water requirements is confounded in areas with lower rainfall those that are presently arid or semiarid. Therefore, climate change will significantly impact agriculture by

S.A. Abd El-Hafez (✉) · A.Z. El-Bably
Water Requirement and Field Irrigation Department, , Water and Environment
Research Institute; Agricultural Research Center, Giza, Egypt

© The Author(s) 2016
S. Ouda, *Major Crops and Water Scarcity in Egypt*,
SpringerBriefs in Water Science and Technology,
DOI 10.1007/978-3-319-21771-0_8

117

increasing water demand, limiting crop productivity, and by reducing water availability in areas where irrigation is most needed or has comparative advantage.

From this point, it is expected that adaptation strategies will focus on minimizing the overall production risk. Adaptation needs are uncertain, but can be defined by specific prediction of likely climate impacts in a specific context. In practice, continued refinement of soil, water, and crop management will contribute to much of the necessary adaptation except in what are already water-stressed conditions. The options for adaptation include the following:

- **Improved agronomic practices**

Cultivation of wheat, maize, or rice on raised bed (wide furrow) reduces yield vulnerability to climate change and yield losses, as a result of improvement in field growing conditions for these crops. Furthermore, adapting that raised bed cultivation under climate change by farmers could be useful for reducing production–consumption gap.

- **Irrigation system**

Adapting sprinkler or drip system under climate change by farmers could be fruitful for reducing production–consumption gap for wheat and maize. Growing wheat and maize under sprinkler and drip system reduces its vulnerability to climate change and reduced yield losses. Water requirements under raised beds and sprinkler system are 20 and 25 % lower than its counterpart under surface irrigation for wheat and maize, respectively. Improving surface irrigation using gated pipes in sugarcane should be considered as an adaptation strategy to reduce climate change risk on sugarcane production. Moreover, using gated pipes as a method for developing surface irrigation in sugarcane increases land productivity under climate change, compared to its counterpart under surface irrigation.

Planting rice on wide furrows as one of the surface irrigation methods is recommended because it saves a big amount of irrigation water that we can use to grow another crop in the same season, like maize, to contribute to increase food security. Wide furrows under current climate have the highest production and the highest water productivity. Also, under climate changes in 2020, 2030, and 2040 recommend wide furrows for the same reasons.

- **Intercropping systems**

Wheat potential production under climate change will be lower by 18 % under relay intercropping of cotton than sole wheat production by 28 % under climate change. So, intercropping cotton on wheat will reduce yield losses climate change, compared to sole wheat production under surface irrigation, raised beds, and sprinkler system.

Summer oil crops can be intercropped with sugarcane, which will increase land productivity and reduce production–consumption gap in edible oil in Egypt that was estimated by 97 %. Furthermore, the intercropped crop will obtain its water requirements from the applied amount of irrigation water for sugarcane, which will increase water productivity.

System level adaptation will respond to strategic policy at national level. Farmers are likely to be highly innovative and proactive in adapting to climate constraints. Therefore, a good understanding of what they do will be required both to match system service to their needs and to assist in broader adoption and dissemination of beneficial practices across irrigation schemes catchments and basins.

- **Crop rotation**

In response to climate change, total water requirements are increasing with climate change in 2040 for the prevailing and proposed crop rotation in old and new lands as well as salt-affected soils in Egypt. But, the rate of increment was higher in the prevailing crop rotation than the proposed crop rotation. Therefore, if we apply the proposed crop rotation, the saved irrigation amount per hectare could be significant in Nile Delta, Middle, and Upper Egypt.

Adopting the proposed crop rotation could reduce the applied water under both current climate and climate change. Crop rotation and intercropping can help in solving food insecurity problem through increase land productivity. Furthermore, implementing crop rotation and intercropping can save sum of irrigation water and increase water productivity. Under climate change in 2040, these saved amounts can be used to maintain the cultivated area that is expected to be reduced as a result of the increase in water requirements of the cultivated crops.

Farmers' Perspectives in Adapting to Climate Change

It is important to link between strategic, system, and farm level development. Although farmers will intuitively adapt to climate trends and more extreme variability, traditional knowledge that has served well for governorates may lose its edge. Amid the scientific excitement of climate change, we should not forget farmers' daily realities and points of view.

As water becomes increasingly scarce, and more expensive, it would be logical to specialize and intensify production to increase returns ($ productivity), although at the cost of greater year to year risk and higher capital investment. Further, the long-term risk associated with capital investment will also increase. Insurance can hedge risks against extreme failures, but is less likely to protect farmers from a generally more extreme climate. Engineered approaches to limiting crop water demand and heat stress will only be afforded by the better-off and more commercially oriented farmers. Those who do intensify are likely to require more secure water supplies, and will need some form of high security water right.

The poorest subsistence farmers will face tough pressures to produce more, in more adverse conditions, with limited capital resources. At the same time, they will be expected to manage their production in a more environmentally sensitive way. Widespread and sustained adoption of drought tolerant and other improved crop varieties will be enhanced if farmers are able to provide their own seeds, and not become locked into buying seeds every season. Dry land farmers and irrigators will require better use of irrigation water in winter and summer crops seasons, such as wheat, maize, rice, and sugarcane and access to better information to adjust management practices, i.e., cultivation on raised beds, intercropping and increasing water application efficiency using modern irrigation systems and gated pipes on the selected crops under current climate, where it can used as adaptations to climate change to reduce climate change risk.

The prospective impacts of climate change on water and implications for irrigation, the impacts of these drivers are likely to include the following:

- Reduction in crop yield and agricultural productivity where temperature constrains crop development. As temperature rises, the efficiency of photosynthesis increases to a maximum and then falls, while the rate of respiration continues to increase more or less up to the point that a plant dies. All other things being equal, the productivity of vegetation thus declines once temperature exceeds an optimum. In general, plants are more sensitive to heat stress at specific (early) stages of growth, (sometimes over relatively short periods) than to seasonal average temperatures. Coupled with increased rates of evapotranspiration, the potential yield and water productivity of crops will fall. However, because yields and water productivity are now low in many parts of Egypt, this does not necessarily mean that they will decline in the long term. Rather, farmers will have to make agronomic improvements to increase productivity from current levels;
- Reduced availability of water in regions in lower, middle, and upper regions; and
- Generally increased evaporative demand from crops as a result of the incremental temperature.

Egypt should be supported in assisting with the development of farmer-oriented programs, which in this case might include explaining climate change and predictive methods to farmers to illustrate the connection between climate change and work that needs to be done to develop a better and more complete understanding of the productivity under climate change through enhancing productivity for major crops and reducing irrigation water; combating adverse sequences of climate change; intercropping oil crops with spring sugarcane to production–consumption gap such as soybean, sesame, and sunflower. These crops will get its water requirements from the applied water to sugarcane. Thus, the cultivated areas of these crops will increase, as well as its national production. So, this procedure can contribute in reduction of edible oil gap exists in Egypt.

It is widely recognized that the transaction costs of monitoring small-scale projects and subsistence farmers exceed the value of benefits, so good incentives for subsistence farmers will be required. As for government incentives, obligatory crop selection and rotation—possibility to introduce obligatory set crop rotation per area in response to water scarcity and climate change impacts, including incentives such as set higher prices for participants, subsidies on fertilizers or assistance with crop marketing, providing security for farmers. This solution would limit farmers' freedom on crop choices, however.

There will be strong pressures for individual farmers to harvest more water by planting intercropping and crop rotation programs. This should happen spontaneously but will also be promoted through soil and water conservation (watershed development) programs. There will be costs in terms of reduced downstream flows (catchment yield) in both surface water, and these apply to more distant farmers.

Planning Adaptation Strategies

Irrigation in a strategic planning context must consider risk; food security; food type; balance of water demands and environmental impacts; substitutability with irrigated agriculture and associated environmental trade-offs. It is timely to re-evaluate the strategic role of irrigation in

- drought proofing of staple crops;
- high-value agriculture, with particular consideration of urban demand and changing food preferences;
- high nutrition value agriculture targeted at subsistence farmers and the poor;
- export earnings versus import substitution; and
- minimizing ecological and climate change sensitive impacts.

The target is to develop an appropriate investment plan for climate sensitive development that is based on future agricultural performance and the probable availability of water resources (in the form of rain, stream flow, surface water, and groundwater storage).

More focused local analysis can be undertaken to better understand adaptation and mitigation options, as well as supporting strategic planning options.

The impacts of climate change on rice production will affect the lives of millions of farmers and consumers. Some of the considerations for adapting rice production systems and incorporating mitigation are listed below:

- Assess reduction likely in paddy area due to reduced runoff effects;
- Determine core paddy areas under climate change scenarios;
- Assess areas that will remain under rice in the summer season;
- Develop methodologies to assess consequent likely reduction in methane emissions and, where possible, incorporate soils information into remote-sensing-based estimates of rice area;
- Differentiate yield impacts across a new rice verities and landscape; and
- Determine rice irrigation strategies and crop diversification strategies to suit.

Development planning for specific region should also attempt better quantification of how mitigation can be achieved in irrigated agriculture through minimal input use (mostly N-fertilizer). Innovative thinking is required to encourage integrated farming systems that combine full irrigation with deficit irrigation production with short duration–high yielding cultivars and agricultural practices.

It is important that planners in Egypt develop the capacity and have access to the tools to undertake to address, and to shore up the information base for decisions particularly with respect to actual water use and current resource availability.

Cooperation Between International Organizations and Development Partners

A number of policy makers presented opportunities for collaboration and concerted development in relation to climate change, food security, poverty alleviation, and economic transitions. The challenge will be to make development assistance climate smart across as well as a wide range of interested parties, who may often have more time-bound agendas.

Practical research on agricultural adaptation through cropping systems research and through crop breeding and testing is required, as are partnerships between the CGIAR centers and research units with established capacity in global and regional climate modeling to evaluate and test resource constraints and options. Such research support and capacity building should be aimed at mainstreaming climate-smart development into local agencies and policy.

Addressing Identified Knowledge Gaps

In Egypt, the agriculture needs for a comprehensive assessment of climate change impacts on agriculture and food security are identified, resulting in the elaboration of adaptive strategies, for different scales and scenarios.

Fundamental and applied research is required to develop effective practices for agricultural methods, irrigation systems, and empirical methods in dry areas and salt-affected soils.

Finally, the impact of climate change on major crops in different situations would help in planning mitigation and adaptation strategies and in finding an appropriate, productive, and economically optimal balance. The goal of synergy between adaptation, mitigation, and sustainable development strategies requires an analytical framework with a sound economic basis.

International Support to Adaptive Strategies

It has identified weaknesses in the existing information bases, the institutional arrangements to oversee water resources management, and the sustainable provision of water to agriculture.

At scheme operational level, there is broad scope to apply management software to calculate water requirements. National and international specialist agencies can promote better understanding of the water resources and agriculture implications of climate change and assist developing countries to improve regional and local projections of impacts in order to develop planned adaptive strategies.

Advocacy would lead on the integration of climate science with agricultural water management and include a strong focus on the preservation and enhancement of natural ecosystems, which are tightly bound to the development and management of irrigated agriculture. This will see further development of an integrated perspective at river basin level, and also across a spectrum of irrigated agriculture.

Two fundamental issues to resolve are how yields and production are likely to change in the future, and how best to provide concrete examples for the extent to which crop adaptation to higher temperatures is possible. The international climate change literature is pessimistic, predicting significant reductions in yield and production, even with adaptation strategies. It will be important to resolve the potential for increases in productivity against a declining potential due to climate change. A separate strand of effort could therefore be directed to the establishment of a public access database on climate-adapted crop varieties, crop rotation, intercropping, agronomic practices, and irrigation systems as well. Some considerable thought and preparation would need to go into the structure of such a database, and into an easy and accessible means of abstracting relevant data. It would be very useful if the database were validated by some testing and evaluation of the field performance of adapted varieties, directly or from secondary data.

The following actions for improving overall water productivity in agricultural will be considered and taken (policy consideration) to face water scarcity:

Improvement of Irrigation Efficiencies

- Carrying out further horizontal expansion, depending on the availability of additional water.
- Prioritizing efficiency measures in affected areas.
- Continuing "Irrigation Improvement Projects" (IIP and IIIMP)-related activities to rehabilitate water distribution systems in prioritized areas, i.e., areas, where drainage water would otherwise flow to sinks and where reuse of drainage water is not recommended because of its adverse effects.
- Providing irrigation advisory services that would include all new development areas.
- Lining in stretches of canals which suffer from high leakage losses.
- Using laser land leveling where possible and where needed to increase filed application efficiencies.
- Controlling drainage during the cultivation of rice.
- Using modern irrigation techniques in all new development areas with light-textured soils.

- Gradually in traducing modern irrigation techniques to replace traditional irrigation methods oases and gradually phasing out the cultivation of rice in these areas.
- Controlling well discharge in desert areas.
- Improving operations and maintenance activities through private participation (water boards and water user association).
- Reducing the irrigation supply after rainfall combined with extra storage upstream from barrage in the Delta.

Improvement of Drainage Conditions

- Continuing the subsurface drainage program of EPADP, with the intention of integrating activities with IIP into IIIMP.

Review of the Drainage Water Reuse Policy in Egypt

- Implementing intermediate reuse at appropriate locations.
- Prioritizing drainage water reuse in areas where:

  - Drainage water would otherwise flow to sinks,
  - The least harm is done to other downstream users, and
  - Groundwater in least vulnerable to pollution.

- Allowing higher permissible salinity of irrigation water after mixing with drainage water.
- Promoting the use of crops that are less sensitive to salinity.

Research and Development Activities Needed

- Update data on natural resources (land, water, and climate) using GIS and satellite imagery data.
- Study integrated agro-climate system and develop the existing agricultural systems (irrigated agriculture, rain fed agriculture, and rangelands)
- Implement integrated main field crops management.
- Implement integrated main horticulture management.

Further Elaborations Are also Needed for the Following Technical Aspects

- Water—use efficiency and water productivity.
- Crop water requirements and irrigation scheduling based on remote-sensing techniques.
- Soil—plant—atmosphere relationships and crop growth modeling.
- Saline and marginal—quality water irrigation climate variability and changes and their impacts on agriculture.
- Land evaluation and agro-ecological performance assessment; operational analysis and rehabilitation.
- Energy consumption in delivering water to where it needed; irrigation water supply and pumping station.
- Water resources management: reservoir operation and groundwater exploitation.

General Recommendations

Expanding the use of modern field irrigation system

- Implementing a mass media campaign to popularize modern field irrigation systems, as well as the support and incentives given by the garment for this purpose, in addition to extension campaigns to achieve this goal.
- Strengthening research in the field of planning and designing modern irrigation systems recommended for each crop and each environment, and the application of and in desert lands.

Enhance the protected agriculture technique

- Implementing an information campaign at the level of priority areas for protected agriculture;
- Informing farmers of the modern techniques used in protected agriculture as a substitute for conventional agriculture in order to save the cost of agriculture inputs;
- Strengthening research in the different fields of protected agriculture, soil-less agriculture, and providing information on the use of information technology and computer-based expert systems; and
- Developing intensified programs for development of human resources working in the field of protected agriculture.

Improving water use efficiency in rain fed agriculture

- Paying greater attention to rain water harvesting projects and expending such projects in accordance with modern techniques;
- Applying supplementary irrigation systems, making use of the results of local and international research;
- Developing infrastructure and facilitates in rain fed areas;
- Developing amps of water basins, in order to raise their use in supplementary irrigation; and
- Media coverage of rain fed areas under development programs, social studies, extension and increased awareness programs.

Rationalizing water use efficiency of groundwater resources

- Supporting and updating detailed studies on groundwater resources, within the framework of developing a national map for groundwater resources, within the framework of developing a national map for groundwater resources, under a national plan to achieve the optimum and sustainable use of such resources;
- Applying modern techniques in monitoring and evaluation grand water and assessing projects and activities to be based on them; and
- Establishing a notional institutional entity to survey, monitor, and manage groundwater laws and establish strict regulations to apply such laws and controls.

Develop unconventional water resources

- Expanding modern projects for the treatment of sewage water through the use of modern techniques;
- Appling on effective auricular and environmental management for the use of nonconventional water Sources (sewage and agricultural drainage water);
- Strengthening research and applied programs in the field of developing and selecting plant varieties that can be grown using saline and low-quality water; and
- Participating in regional and international research efforts cost efficient seawater desalinization.